AF327365

CELL BIOLOGY RESEARCH PROGRESS

# PROGENITOR CELLS

# BIOLOGY, CHARACTERIZATION AND POTENTIAL CLINICAL APPLICATIONS

# CELL BIOLOGY RESEARCH PROGRESS

Additional books in this series can be found on Nova's website
under the Series tab.


Additional e-books in this series can be found on Nova's website
under the e-book tab.

CELL BIOLOGY RESEARCH PROGRESS

# PROGENITOR CELLS

# BIOLOGY, CHARACTERIZATION AND POTENTIAL CLINICAL APPLICATIONS

PATRICK M. HORTON

AND

BRETT E. LAWRENCE

EDITORS

*New York*

For permission to use material from this book please contact us:
Telephone 631-231-7269; Fax 631-231-8175
Web Site: http://www.novapublishers.com

## NOTICE TO THE READER



Additional color graphics may be available in the e-book version of this book.

**Library of Congress Cataloging-in-Publication Data**

ISBN: 978-1-62808-994-3

Library of Congress Control Number: 2013948183

*Published by Nova Science Publishers, Inc. † New York*

# Contents

# Preface

In this book, the authors present current research in the study of the biology, characterization and potential clinical application of progenitor cells. Topics discussed include endothelial progenitor cells and cardiovascular disease; isolation, expansion and clinical therapy of human corneal epithelial stem/progenitor cells; genetically engineered blood pharming; regulation of neural progenitor cells by Wnt5a-signaling in the developmental central nervous system; endothelial progenitor cells in a clinical setting; role of microRNAs in endothelial progenitor cells and their implication for cardiac repair; and cellular origins in amphibian regeneration.

Chapter 1 - Endothelial Progenitor Cells (EPCs) are bone-marrow derived stem cells that are postulated to contribute to post-natal vasculogenesis and to repair of damaged endothelium by incorporation into the vessel wall, secretion of paracrine hormones and stimulation of angiogenesis. Since the first description of the putative EPCs in 1997, and the role of these cells in neovascularisation of mouse and rabbit ischaemic limbs was originally described, there has been an explosion of research into the role of EPCs in human cardiovascular disease. There is now a large body of direct and indirect evidence to support an important role for EPCs in cardiovascular disease processes. This book chapter explores the following:

1. Correlation between EPCs and other cardiovascular risk markers
2. EPCs in patients with established cardiovascular disease
3. Reversible defects in EPC number and function in patients with an increased cardiovascular risk
4. Statins and EPC biology

5. The effect on EPCs of other interventions known to reduce cardiovascular risk - EPCs and treatment of diabetes, hypertension, subclinical hypothyroidism
6. Beneficial effects of EPC-based therapies animal models of ischaemia
7. Human Studies of EPC-based therapies

A lower level of circulating EPCs and reduced EPC function in vitro are associated with an increased cardiovascular risk. The accumulated evidence suggests that a balance between the damaging effects of conventional cardiovascular risk factors and the ability of circulating EPCs to effect endothelial repair determines this cardiovascular risk.

Chapter 2 - Stem cells can be defined as cells that have the capacity to self-renew and the ability to generate differentiated progeny or multiple cell lineages. True stem cells can turn into any type of cells, while progenitor cells are more or less committed to becoming cell types of a particular tissue. Human corneal epithelial stem cells (CESCs) represent a great example and model of adult stem or progenitor cells. Human CESCs have been identified to locate in the basal epithelial layer of the limbus, and thus also referred as to limbal stem cells. The authors would like to use the both terms, stem and progenitor cells in this chapter based on previous use in the literature for more than two decades. Although the CESCs have been identified to reside at the limbus and many stem cell markers have been proposed, there is no consensus to date regarding the definitive markers for CESCs, and identification and isolation of these cells are still challenging. Based on evaluation of a variety of proposed markers, the authors have characterized that the CESCs located in the basal layer of human limbal epithelium are small primitive cells expressing three patterns of molecular markers, which represent a unique phenotype of putative corneal epithelial stem or progenitor cells.

Based on adult stem cell criteria and the putative limbal stem cell phenotype, the authors' group has attempted to enrich for human CESCs through novel approaches including cell-sizing, adhering to extracellular matrix collagen type IV, and cell sorting for side population or for expression of ABCG2 or connexin 43 cell surface markers. The 5 clonogenic populations isolated from limbal epithelium and its cultures by different methods show the properties that are characteristics of adult stem/progenitor cells: 1) relatively undifferentiated, 2) high proliferative potential, 3) self-renewal. Expansion and cultivation of corneal epithelial progenitor cells have been achieved using different methods, such as limbal tissue explant culture, and limbal epithelial cell suspension co-culture with mouse 3T3 fibroblast feed layer. To avoid the use of xeno-components, two cell lines of commercial human fibroblasts have been identified that support human corneal epithelial regeneration, and have potential use in replacing mouse 3T3 cells for corneal tissue bioengineering.

The concept of CESCs has formed the basis for identifying a class of blinding diseases that display features of corneal epithelial stem cell deficiency or limbal stem cell deficiency (LSCD), where the limbal epithelium is damaged. LSCD is characterized by persistent or recurrent epithelial defects, ulceration, corneal vascularization, chronic inflammation, scarring, and conjunctivalization (conjunctival epithelial ingrowth). Only transplantation of CESCs can restore vision. Due to an increasing shortage of corneal donors, corneal tissue engineering is becoming an important discipline that holds great promise for corneal reconstruction. CESCs and optical substrates are known to be the most important factors for corneal tissue bioengineering in regenerative medicine. The authors' team has recently explored the utilization of natural donor corneal stroma in corneal tissue engineering. In combination with fresh limbal epithelium containing stem cells, and the donor corneal stroma, a great source of natural optical substrate, the authors developed a native-like corneal equivalent construct with proliferative potential. This corneal construct provides a new clinical cell therapy for corneal reconstruction.

Chapter 3 - Blood pharming is a recently designed concept to enable in vitro production of blood cells that are safe, effective and readily available. This approach represents an alternative to blood donation and may contribute to overcome the shortage of blood products. However, the high variability of the human leukocyte antigen (HLA) loci remains a major hurdle to the application of off-the-shelf blood products. Refractoriness to platelet (PLT) transfusion caused by alloimmunization against HLA class I antigens constitutes a relevant clinical problem. Thus, it would be desirable to generate PLT units devoid of HLA antigens. To reduce the immunogenicity of cell-based therapeutics, the authors have permanently reduced HLA class I expression using an RNA interference strategy. Furthermore, they demonstrated that the generation of HLA class I-silenced (HLA-universal) PLTs from CD34+ progenitor cells using an shRNA targeting $\beta$2-microglobulin transcripts is feasible. CD34+ progenitor cells derived from G-CSF mobilised donors were transduced with a lentiviral vector encoding for the $\beta$2-microglobulin-specific shRNA and differentiated into PLTs using a liquid culture system. The functionality of HLA-silenced PLTs and their ability to escape HLA antibody-mediated cytotoxicity were evaluated *in vitro* and *in vivo*. Platelet activation in response to ADP and thrombin were assessed *in vitro*. The immune-evasion capability of HLA-universal megakaryocytes

(MKs) and PLTs was tested in lymphocytotoxicity assays using anti-HLA antibodies. To assess the functionality of HLA-universal PLTs *in vivo*, HLA-silenced MKs were infused into NOD/SCID/IL-2Rγc$^{-/-}$ mice with or without anti-HLA antibodies. PLT generation was evaluated by flow cytometry using anti-CD42a and CD61 antibodies. HLA-universal PLTs demonstrated to be functionally similar to blood-derived PLTs. Lymphocytotoxicity assays showed that HLA-silencing efficiently protects MKs against HLA antibody-mediated complement-dependent cytotoxicity. 80-90% of HLA-expressing MKs, but only 3% of HLA-silenced MKs were lysed. *In vivo*, both HLA-expressing and HLA-silenced MKs showed human PLT production (up to 0.5% within the PLT population) when anti-HLA antibodies were absent. However, in presence of anti-HLA antibodies HLA-expressing MKs were rapidly cleared from the circulation of mice, while HLA-silenced MKs escaped HLA antibody-mediated cytotoxicity and human PLT production was detectable up to 11 days. The authors' studies show that HLA-silenced PLTs are functional and efficiently protected against HLA antibody-mediated cytotoxicity. In this chapter, the authors provide a review of their most recent findings in the use of CD34+ progenitor cells for the production of HLA-universal PLTs and their potential clinical applications. Provision of HLA-universal PLT units may become an important component in the management of patients with PLT transfusion refractoriness.

Chapter 4 - In the developing central nervous system (CNS), many different types of neurons are generated in an age- and brain region-dependent manner. Thus, developmental neurogenesis must be precisely controlled in order to ensure the well-ordered generation of the appropriate numbers of specialized neurons. Neurons are generally produced from neural progenitor cells (NPCs). Recent findings have indicated that neuronal subtypes are determined partly in NPCs prior to their differentiation. In addition, the number of neurons generated from NPCs mainly depends on a balance between proliferation and differentiation of NPCs, which is regulated by both intrinsic and extrinsic factors. The Wnt-family of secreted glycoproteins, well documented extrinsic factors in the CNS, have been shown to play crucial roles in regulating developmental neurogenesis. Wnt signaling can be largely classified into β-catenin-dependent (canonical) and -independent (non-canonical) pathways, which are mediated by binding of different Wnt ligands and their cognate receptors, including Frizzled (Fzd), LRP5/6, Ryk and Ror1/2.

In this chapter, the authors focus on the function of Wnt5a-induced signaling in the regulation of NPCs during developmental neurogenesis.

Wnt5a can elicit non-canonical Wnt signaling to regulate diverse cellular functions. The authors have recently shown that Wnt5a acts to enhance neurogenesis in the cerebral cortex through the maintenance of proliferative and neurogenic NPCs through its binding to the Ror-family of receptor tyrosine kinases, Ror1 and Ror2. In the developing cerebral cortex, glutamatergic neurons are generated from NPCs in the ventricular zone through the production of intermediate progenitor cells (IPCs), which can divide prior to differentiation into neurons to expand the population of neurons. Inhibition of Wnt5a-Ror signaling in NPCs results in a decreased number of IPCs, eventually leading to the reduction of total number of neurons generated from NPCs. In the developing midbrain and olfactory bulb, Wnt5a has been shown to promote the differentiation of NPCs into dopaminergic neurons or GABAergic interneurons, respectively. In both cases, Wnt5a appears to act to specify the differentiation of NPCs into a distinct subtype of neurons, because Wnt5a stimulation increases the number of dopaminergic or GABAergic neurons rather than the total number of neurons. Taken together, these findings indicate that Wnt5a promotes the production of specific neuron subtypes possibly through the regulation of NPCs in different manners, depending on the brain regions. Importantly, the ability of Wnt5a-signaling to provide specific neuronal populations from NPCs as renewable sources would be beneficial for producing replacement cells for therapeutic applications to neurological disorders, i.e. glutamatergic neurons for stroke, dopaminergic neurons for Parkinson's disease, and GABAergic neurons for Huntington's disease.

Chapter 5 - Senescence of cells is associated with shortened or damaged telomeres and is characterized by permanent exit from the cell cycle and altered function. Cellular senescence is caused by repeated cell division, and also conditions of stress including inflammation and reactive oxygen species can lead to the development of premature senescence. At the cellular level, proliferative and oxidative-stress induced cell senescence related to a pro-inflammatory state might strongly contribute to age-associated impaired tissue and organ functions. Vascular cells (endothelial cells, vascular smooth muscle cells) and bone marrow-derived endothelial progenitor cells have been repeatedly shown to have pivotal role in the maintenance and regeneration of cardiovascular tissue. Therefore, the molecular mechanisms of vascular cell senescence have been extensively studied. However, therapeutic approaches to prevent cellular senescence in cardiovascular disease (CVD) are still limited.

Hepatocyte growth factor (HGF), vascular endothelial growth factor (VEGF), and fibroblast growth factor (FGF) are all potent angiogenic growth factors in animal models of ischemia, but their therapeutic effects are not the same in animal experiments and clinical trials. A multicenter, double-blind, placebo-controlled phase III clinical trial in Japan and a US phase II clinical trial of HGF gene therapy for critical limb ischemia (CLI) demonstrated a significant improvement in primary end points and an increase in transcutaneous partial pressure of oxygen even after one year compared with placebo, whereas effectiveness of VEGF and FGF treatment for CLI has not yet been shown. Moreover, the authors' recent publication and another researcher demonstrated that HGF acts as an anti-inflammatory cytokine, while VEGF and FGF act as pro-inflammatory cytokine.

This review overviews the outcomes of clinical trials using angiogenic growth factors, which have shown a dramatic effect in several animal studies. Additionally, interventions with HGF aimed at improving the regenerative capacity of stem/progenitor cells and vascular cells by preventing cellular senescence are discussed.

Chapter 6 - Endothelial progenitor cells (EPC) are mobilized after myocardial infarction (MI) from the bone marrow to injured sites of the heart where they participate in cardiac repair by revascularization of ischemic tissues. Endothelial progenitor cells have been actively studied, but their exact phenotype and regenerative properties are still controversial. Small trials with progenitor cells of different origins showed modest clinical benefits. It is assumed that a better understanding of the biology of EPC will contribute to improve their therapeutic potential. MicroRNAs (miRNAs) are small single-stranded non-coding RNAs that modulate gene expression by interacting post transcriptionally with protein-coding RNAs. MicroRNAs regulate multiple biological processes involved in cardiac development and disease. While many studies addressed the role of miRNAs in cardiac cells, less is known of the effect of miRNAs in EPC. Recent studies showed that miRNAs indeed regulate the biology of EPC. Since novel technologies to enhance or blunt the functions of miRNAs have been recently developed, it is conceivable that miRNAs may become promising new therapeutic tools. This article will review the recent advances in the knowledge of the effects of miRNAs in EPC and will discuss how miRNAs could be manipulated to improve the regenerative capacities of EPC in the diseased heart.

Chapter 7 - Vertebrate animals utilize a diversity of ways to produce cells for regenerating missing body part. In addition to dedifferentiation and transdifferentiation, recent studies in amphibian appendage regeneration have demonstrated the importance of tissue specific progenitor cells in reconstitution of regenerated tissues after amputation or injury. Accumulating evidence supports the notion that a progenitor-based tissue regeneration mechanism is common in vertebrate animals, including mammals. In this chapter, the authors will examine the cellular origin during amphibian regeneration, and use *Xenopus* limb as an example to discuss the usefulness of progenitor cells for possible measures in stimulating regeneration.

In: Progenitor Cells                                   ISBN: 978-1-62808-994-3
Editors: P. M. Horton, B. E. Lawrence  © 2013 Nova Science Publishers, Inc.

*Chapter 1*

---

# Endothelial Progenitor Cells and Cardiovascular Disease

---

***Thomas F. J. King and John H. McDermott***
Department of Endocrinology and Diabetes, Royal College of
Surgeons in Ireland, Connolly Hospital, Dublin, Ireland

## Abstract

Endothelial Progenitor Cells (EPCs) are bone-marrow derived stem
cells that are postulated to contribute to post-natal vasculogenesis and to
repair of damaged endothelium by incorporation into the vessel wall,
secretion of paracrine hormones and stimulation of angiogenesis. Since
the first description of the putative EPCs in 1997, and the role of these
cells in neovascularisation of mouse and rabbit ischaemic limbs was
originally described, there has been an explosion of research into the role
of EPCs in human cardiovascular disease. There is now a large body of
direct and indirect evidence to support an important role for EPCs in
cardiovascular disease processes. This book chapter explores the
following:

1. Correlation between EPCs and other cardiovascular risk
   markers
2. EPCs in patients with established cardiovascular disease
3. Reversible defects in EPC number and function in patients with
   an increased cardiovascular risk
4. Statins and EPC biology

5.　The effect on EPCs of other interventions known to reduce cardiovascular risk - EPCs and treatment of diabetes, hypertension, subclinical hypothyroidism
6.　Beneficial effects of EPC-based therapies animal models of ischaemia
7.　Human Studies of EPC-based therapies

A lower level of circulating EPCs and reduced EPC function in vitro are associated with an increased cardiovascular risk. The accumulated evidence suggests that a balance between the damaging effects of conventional cardiovascular risk factors and the ability of circulating EPCs to effect endothelial repair determines this cardiovascular risk.

# 1. Introduction

With the evolution from single cell to multi-cellular organisms came the need to supply a large number of cells at proportionately great physical separation with the necessary nutrients for survival. Blood provides such nutrients, with blood vessels, lined by endothelial cells, carrying the blood to every cell in the body. As well as acting as conduits for blood cells, the endothelial cells lining blood vessels perform numerous other functions vital to health, however: acting as a protective, selective barrier between circulating blood and tissues; detecting changes in haemodynamic forces and responding by releasing autocrine and paracrine factors to maintain vascular homeostasis; and preventing thrombosis in normal conditions, but promoting thrombosis when appropriate [1]. It is clear, therefore, that a healthy, fully-functioning vasculature is vital to the health of the entire organism, and the important role of the endothelium in health and disease has become increasingly apparent in recent years.

The process of blood vessel formation *in utero* proceeds by vasculogenesis, whereby undifferentiated angioblasts sprout vessels that invade and develop in parallel with the developing organs – angioblasts also differentiate into blood cells, meaning that the endothelium and the circulating blood cells share a common precursor [2]. In the adult, however, new blood vessel formation and repair of damaged vessels in the setting of endothelial dysfunction was, by convention, thought to occur by the processes of angiogenesis and arteriogenesis, both of which processes involve mature endothelial cells already *in situ* in the blood vessel wall. Such cells by their nature possess limited regenerative capacity, particularly if damaged by oxidative stresses, hypertension or other insults, including, of course, the aging process [3]. The exciting concept that an adult equivalent of the haemangioblast could exist was initially prompted by reports of the endothelialization of prosthetic grafts by blood-derived cells[4, 5]. Scott et al. [4] suspended pledgets of vascular graft material within the aorta of dogs using metal stents- the material was not, therefore, in contact with the surface of the aorta. Despite this, endothelial cells were discovered coating the graft material at 55 days post-implantation, strongly suggesting endothelialization by circulating (and therefore probably bone marrow derived) cells, rather than by cells from the aorta itself. Rafii et al. [6] investigated the composition of the non-thrombotic neointimal surface observed on the surface of implantable left ventricular assist devices, and found spindle-shaped and CD34-positive cells, which proliferated in culture when plated, among a variety of cells in the neointima. These studies pointed, in theory, to the existence of a circulating adult haemangioblast, the putative endothelial progenitor cell - but isolation of the cell itself remained elusive for a number of years.

The subsequent discovery of endothelial progenitor cells in 1997 [7] heralded an explosion of research in the area, based on the possible vast therapeutic potential for these cells: a previously undifferentiated and therefore undamaged cell, with the ability to mobilize from the bone-marrow, to home to sites of vascular damage, to incorporate into damaged and dysfunctional endothelium, and to repair damaged or dysfunctional blood vessels or to participate in new blood vessel formation. The potential benefits in patients with endothelial dysfunction, and cardiovascular or peripheral vascular disease were immediately obvious.

In their seminal paper describing the first successful isolation of EPCs, Asahara et al. isolated CD34 positive mononuclear cells from human peripheral blood using CD34-coated magnetic beads. Isolated mononuclear cells were plated on plastic, fibronectin, or collagen type 1. Cells plated on fibronectin attached and became spindle-shaped within 3 days. The authors next labelled similarly isolated CD34-positive cells with DiI and co-plated the labelled cells with unlabelled CD34-negative cells on fibronectin. After 7 days, 60% of cells attached were DiI positive, even though CD34-positive cells accounted for less than 1% of the originally plated cells. This co-culture of CD34-positive with CD34 –negative cells increased the proliferation rate of CD34-positive cells. The authors confirmed the endothelial phenotype of the CD34-positive cells after 7 days in culture by documenting expression of eNOS and KDR. They next attempted to determine if the cultured CD34-positive cells were, in fact, the putative adult haemangioblast as hypothesised, by assessing the ability of these cells to promote neovascularaiztion. Two days after creating hind-limb ischaemia in athymic nude mice, CD34-positive or negative cells were injected into the mouse tail veins. On post-mortem histological examination 1 to 6 weeks later, numerous DiI labelled cells were present in the neovascularized hind-limb, and almost all cells were incorporated into the capillary wall. 13.4% of capillaries contained DiI-positive cells in mice injected with CD34-positive cells, versus 1.6% of capillaries in mice injected with CD34-negative cells. Importantly, no DiI positive cells were found in the uninjured limb.

The paper by Asahara et al., therefore, demonstrated that cells isolated from the mononuclear fraction of peripheral blood, and possessing some of the characteristics of immature/stem cells, could differentiate into cells of an endothelial phenotype under certain culture conditions *in vitro*, and that such cells, when infused intravenously, were capable of selectively homing to areas of ischaemia and participating in new blood vessel formation. Many subsequent studies, detailed in this chapter, have confirmed the importance of the EPC in cardiovascular health and disease.

The question 'what is an EPC?' may seem a strange one given the cogent and logical arguments above. However, despite the ever-increasing number of publications in the area of EPC biology, much of the EPC literature has been dogged by the inability to accurately define an EPC, or to define a universally accepted marker for EPCs. The 'true' or 'ideal' EPC, and the cell originally pursued by Asahara and colleagues, is a cell that:

- originates in the bone marrow and resides in the bone marrow in an undifferentiated, immature state,
- proliferates in an immature state,
- is mobilized into the blood stream in response to blood-borne signals,
- circulates in the blood,
- homes to and incorporates into an area of endothelial damage or new blood vessel formation,
- differentiates *in situ* into a mature endothelial cell,
- contributes to and enhances the processes of endothelial repair and new blood vessel formation

In the laboratory, therefore, a 'true' EPC should be capable of being isolated from the blood or bone marrow, should be capable of proliferation, and should initially display an undifferentiated phenotype capable of developing into an endothelial phenotype in appropriate conditions. Such isolated cells, when infused into the bloodstream, should incorporate into blood vessels and improve endothelial function and ameliorate ischaemia.

The paper by Asahara et al. studied cells that possessed many of these features. It is rare, however, that any two laboratories conducting EPC research have utilized the same conditions to isolate these EPCs. In fact, the method utilized by Asahara et al. in their seminal paper has been largely ignored since their original paper. A particularly stark illustration of the different methodologies employed by different laboratories, and the confusing terminology in the field of EPCs, is the approach to non-adherent cells present in culture at days 2- post-plating. In many studies, including those from our laboratory, non-adherent cells in culture at day 4 are removed and discarded, and culture of adherent cells is continued. In contrast, some groups use a re-plating method whereby the non-adherent cells in culture are removed, re-plated, and cultured to day 7 [8]. These are the very cells that are removed and discarded in our laboratory, and yet these cells are still termed EPCs when cultured to day 7 in other laboratories.

Many recent papers have attempted to deal with some of the criticisms above by utilizing flow-cytometry techniques in an attempt to better define the cells under study [9]. These studies use flow cytometric analysis to count the number or percentage of cells expressing cell-surface antigens of interest. Fluorescently-labelled antibodies against cell-surface antigens representing markers of immaturity, or bone-marrow origin, such as CD34, and against antigens denoting endothelial lineage, such as KDR or Ve-Cadherin, are added to peripheral blood or freshly-isolated monocytes, and analysis performed to determine the number of cells expressing both markers. Cells combining bone marrow origin with endothelial lineage and therefore co-expressing CD34 and KDR, for instance, are deemed to be EPCs by virtue of this co-expression. Such flow-based studies have the advantage of minimising the number of laboratory steps required before achieving a cell count, and therefore, in theory, more accurately reflecting the actual number of circulating cells *in vivo*. Studies utilizing FACS analysis to count the number of circulating EPCs have been performed in a number of disease states and have found a decline in EPCs as measured by these techniques in high-cardiovascular risk states [10-15], and will be mentioned later in this chapter. While FACS-based EPC analysis has certainly demonstrated a link between EPCs and cardiovascular disease, functional analysis of the EPCs under study cannot be performed, nor can it be directly proven that such EPCs, if infused in settings of ischaemia, will have a therapeutic benefit *in vivo*. An interesting paper regarding the different methods of EPC analysis compared 6 different flow cytometry methods and a CFU assay, and found only moderate to poor correlation between different methods [16].

In the midst of the confusion and terminological difficulties that are obvious in the still nascent field of endothelial progenitor cell biology, however, there is a vast amount of incontrovertible evidence regarding the importance of these cells in cardiovascular disease, and the potential manipulation of EPCs for therapeutic benefit remains.

Studies have found:

- A correlation between EPC number, endothelial function and cardiovascular risk factors
- Reduced number of EPCs in patients with risk factors for cardiovascular disease such as metabolic syndrome, diabetes and smoking
- Reduced EPC number and function in patients with critical limb ischaemia and coronary artery disease, and reduced EPC number is a significant independent predictor of poor prognosis in patients with established coronary artery disease
- Interventions that reduce cardiovascular risk such as exercise and weight loss improve EPC number and function
- Reduced EPC number and function in obesity, which both normalise after weight loss, and EPCs from obese subjects show a reduced angiogenic response when injected into an animal ischaemia model
- Reduced EPC number in conditions that increase cardiovascular risk, and subsequent amelioration in EPC number with medical treatment, for example diabetes, hypertension and sub-clinical hypothyroidism
- Statins increase EPC number and function both *in vitro* and *in vivo*
- In animal models, exogenously administered EPCs migrate to areas of ischaemia where they mediate neovascularisation and improve endothelial function
- In human studies exogenous EPC infusions are potentially beneficial in coronary and peripheral arterial disease

The accumulated evidence suggests that a balance between the damaging effects of conventional cardiovascular risk factors and the ability of circulating EPCs to effect endothelial repair determines cardiovascular risk.

# 2. Correlation between EPCs and Other Cardiovascular Risk Markers

Hill et al. [8] cultured EPCs from 45 men with various degrees of cardiovascular risk, but without overt cardiovascular disease, and found a strong correlation between EPC number and combined Framingham risk factor score (r=-0.47, P=0.001). Measurement of flow-mediated brachial-artery reactivity also revealed a significant relation between endothelial function and the number of progenitor cells (r=0.59, P<0.001). Levels of circulating EPCs in this study were a better predictor of vascular reactivity than the presence or absence of conventional cardiovascular risk factors. A β-galactosidase activity assay was used as a marker of cellular senescence on day 7 EPCs, and a higher percentage of EPCs from patients with high Framingham risk scores displayed a senescent phenotype (72±15 percent of the cells derived from the high-risk subjects vs. 27±9 percent of the cells from the low-risk subjects had b-galactosidase staining (P=0.005)). The authors postulated that endothelial injury in the absence of sufficient circulating progenitor cells may affect the progression of cardiovascular disease.

Jialal et al. [17] measured EPCs in subjects with and without metabolic syndrome using flow cytometry (CD34+ KDR+) and cell culture (CFU-EC method). Smoking, atherosclerosis and diabetes were exclusion criteria and Metabolic Syndrome was defined as having 3 out of: central obesity, hypertension, dyslipidaemia or hypertension. The subjects with metabolic syndrome had significantly lower (34%, p<0.001) CD34+KDR+ EPCs compared with age-matched healthy volunteers. There were significantly fewer EPCs in culture in the metabolic syndrome group (48%, p<0.001). Regression analysis revealed that CRP, triglycerides, age, and plasma glucose were the strongest predictors of reduced circulating CD34+KDR+ cells (adjusted R-squared = 0.147, model p = 0.01).

Studies in patients with both type 1 and type 2 diabetes have shown a reduction in EPC numbers compared to their healthy counterparts, and an inverse correlation with EPC number and HbA1c [18-20]. Recent studies from our own laboratory show a reduction in EPC number and function in mothers of low birthweight babies, a group of patients with a known elevated future cardiovascular risk [21].

# 3. EPCs in Patients with Established Cardiovascular Disease

To assess the role of EPCs in patients with established coronary artery disease (CAD), Vasa et al. [22] studied 45 subjects with angiographic evidence of coronary artery disease, and 15 healthy controls. The number of cells adherent to fibronectin after 4 days culture in EBM with dual positive staining for LDL and lectin were counted, flow cytometry was performed to measure circulating levels of CD34+/KDR+ cells, and EPC migratory function was assessed with a modified Boyden chamber (for 24 hours, with VEGF as a chemo attractant). To determine the influence of atherosclerotic risk factors, a risk factor score including age, sex, hypertension, diabetes, smoking, positive family history of CAD, and LDL cholesterol levels was used. In this study the number of cardiovascular risk factors was significantly negatively correlated numbers of cultured EPCs (R=-0.394, P=0.002) and CD34+/KDR+ cells (R=-0.537, P<0.001). Analysis of the individual risk factors demonstrated that smokers had significantly reduced levels of cultured EPCs (P<0.001) and CD34+/KDR+ cells (P=0.003). EPCs isolated from patients with CAD also revealed an impaired migratory response, which was inversely correlated with the number of risk factors (R=-0.484, P=0.002). By multivariate analysis, hypertension was identified as a major independent predictor for impaired EPC migration (P=0.043). The authors concluded that the correlation with risk factors, and the reduced levels and functional impairment of EPCs observed, may contribute to impaired vascularisation in patients with CAD.

Yue et al. [23] performed a cross sectional observational study of 174 patients with established CAD. They found significantly lower circulating log CD34/KDR(+) EPCs in smokers compared with non-smokers (0.86 +/- 0.03 vs 0.96 +/- 0.03 x $10^{-3}$/ml, p = 0.032). Smokers with elevated pulmonary artery systolic pressure (PASP) also had significantly lower circulating EPCs, higher pulmonary vascular resistance, and larger right ventricular dimensions with impaired function. Log CD34/KDR+ and log CD133/KDR+ EPC counts were significantly and negatively correlated with PASP (r = −0.30, p <0.001, and r = −0.34, p <0.001, respectively) and pulmonary vascular resistance (r = −0.29, p = 0.002, and r = −0.18, p = 0.013, respectively). The reduced number of circulating EPCs and elevated PASP in smokers with CAD suggests that in smokers, depletion of circulating EPCs might be linked to the occurrence of pulmonary vascular dysfunction.

Schmidt-Lucke et al. [24] measured circulating EPCs (defined as CD34+ KDR+ cells on flow cytometry) in 120 patients with acute coronary syndrome, stable coronary artery disease and control subjects. Patients were followed up for a median 10-month period and the primary outcome was cardiovascular events (cardiovascular death, unstable angina, myocardial infarction, PTCA, CABG, or ischaemic stroke). The authors found that patients suffering from cardiovascular events had significantly lower numbers of EPCs (P<0.05). Reduced numbers of EPCs were associated with a significantly higher incidence of cardiovascular events by Kaplan-Meier analysis (P=0.0009). By multivariate analysis, a reduced number of EPCs was a significant, independent predictor of poor prognosis, even after adjustment for traditional cardiovascular risk factors and disease activity (hazard ratio, 3.9; P<0.05).

Werner et al. [25] measured CD34+ KDR+ cells with flow cytometry in 519 patients with coronary artery disease as confirmed on angiography, and patients were followed up for 12 months. After adjustment for age, sex, vascular risk factors, and other relevant variables, they found that increased levels of endothelial progenitor cells were associated with a reduced risk of death from cardiovascular causes (hazard ratio, 0.31; 95 percent confidence interval, 0.16 to 0.63; P=0.001), a first major cardiovascular event (hazard ratio, 0.74; 95 percent confidence interval, 0.62 to 0.89; P=0.002), a need for revascularization (hazard ratio, 0.77; 95 percent confidence interval, 0.62 to 0.95; P=0.02), and hospitalization (hazard ratio, 0.76; 95 percent confidence interval, 0.63 to 0.94; P=0.01). However, EPC levels were not predictive of myocardial infarction or of death from all causes. The authors concluded that the level of circulating EPCs predicts the occurrence of cardiovascular events and death from cardiovascular causes and may help to identify patients at increased cardiovascular risk.

Chen et al. [26] cultured EPCs from 74 patients with Type 2 diabetes with and without critical leg ischaemia, and non-diabetic patients with and without lower extremity vascular disease. A modified Boyden chamber with VEGF as a chemo-attractant was used to assess cultured EPC migratory function *in vitro*. The migratory function was significantly impaired in diabetic patients without (48 count/view/well) and with (51 count/view/well) critical leg ischaemia and non-diabetic patients with critical leg ischaemia (49 count/view/well) compared with healthy subjects (63 count/view/well), p < 0.0001. As the migratory function of EPCs was impaired in patients with Type 2 diabetes, even in those without critical leg ischaemia, the authors postulated that Type 2 diabetes may alter EPC function and may account for the impaired neovascularisation and more aggressive clinical course in the development of critical limb ischaemia in patients with Type 2 diabetes compared with non-diabetic patients.

Huang et al. [27] demonstrated a reduction in both circulating and cultured EPCs in patients with coronary artery disease compared to controls, and a correlation between homocysteine levels and EPC number.

# 4. Reversible Defects in EPC Number and Function in Patients with an Increased Cardiovascular Risk

The Austrian group mentioned above also conducted a prospective randomised control trial of 40 patients with symptomatic peripheral arterial disease [28]. They measured EPC by flow cytometry (CD34+/KDR+/CD133+ cells were deemed EPCs) and cell culture, and EPC migration was assessed using a modified Boyden Chamber. After 6 months of Supervised Exercise Training, they noted a significant increase in both measurements of EPC number, and migratory activity was also significantly increased.

Heida et al. [29] studied EPC number and function in obese patients before and after weight loss. 49 obese (body mass index 42 +/- 7 kg/m2) and 49 age-matched lean volunteers wre studied. EPC number and function (migration through a modified Boyden chamber, adhesion to fibronectin and angiogenesis with HUVEC/Matrigel assay) were assessed at baseline, and in obese subjects who lost weight after 6 months (defined as current BMI <35 kg/m2 and/or >10% loss of body weight compared with baseline). The authors noted a lower EPC count at baseline in obese subjects, and that the EPCs cultivated from obese subjects displayed impaired adhesion, impaired migratory activity and an impaired ability to incorporate into network-like structures. The investigators went on to study the EPCs *in vivo* using a mouse hind limb ischaemia model. When injected into the mice, labelled EPCs from obese subjects were less frequently detected within the ischaemic hind limb musculature after 10 days (28 +/- 25 vs. 88 +/- 85 chloromethylbenzamido-DiI–positive cells/mm2; p = 0.017; 11 mice per group). They also found that mice treated with EPCs from obese subjects revealed a reduced angiogenic response as assessed by the number of CD31-positive cells per square millimeter (p = 0.036 vs. lean). After 6 months of a weight loss programme, 26 subjects had repeat sampling for EPC number and function. There was no change in medications over the period and mean weight loss was 26kg, with

BMI dropping from 43 to 35 $Kg/m^2$, (p <0.001 for weight and BMI differences from baseline). There was no significant difference in fasting glucose, total cholesterol or rates of hypertension but serum triglycerides were lower (114 vs 149 mg/dl, p = 0.004) as was LDL cholesterol (115 vs 126 mg/dl, p = 0.024) when compared to pre-weight loss values. Weight loss seemed to restore the number of acetylated low-density lipoprotein, lectin double-positive EPCs (p < 0.05 vs. initially obese, 10 per group). Weight loss also improved the adhesive properties of EPCs on fibronectin (p < 0.05 vs. initially obese), normalised the migratory activity of EPCs (p < 0.001 vs. initially obese and p > 0.05 vs. lean; 8 per group), and restored EPCs angiogenic properties in a Matrigel assay (p < 0.01 and p > 0.05, respectively; n = 7). Compared to lean controls, therefore, obese subjects had reduced EPC numbers with reduced adhesive and migratory capacity, but these defects were reversible with significant weight loss.

Patients with primary aldosteronism have been shown to have lower circulating and cultured EPC numbers compared to patients with essential hypertension [30], and high-dose aldosterone has been shown to attenuate EPC proliferation and angiogenesis *in vitro*. When patients were followed up 6 months after surgical resection of the aldosterone producing tumour, EPC counts returned to levels that were comparable to those in the essential hypertension group. Interestingly, the preoperative number of EPCs predicted the curabilityof hypertension after adrenalectomy.

# 5. Statins and EPC Biology

HMG-CoA reductase inhibitors (Statins) inhibit the rate-limiting step in the formation of cholesterol, resulting in a reduction in serum cholesterol levels. In addition they have anti-inflammatory and anti-thrombotic properties, and have been shown to reduce cardiovascular risk independent of their lipid lowering properties [31]. The previously documented links between EPCs and cardiovascular disease and the beneficial effects of statins in cardiovascular disease inevitably has led to a number of studies on the effects of statin therapy on EPCs.

Dimmeler et al. [32] found that addition of atorvastatin to culture media increased the number of differentiated adherent EPCs (dual positive Lectin/DiLDL cells) at day 3 of culture. To test the *in vivo* relevance of this findings, mice were fed with simvastatin (20 mg/kg daily) for 3 weeks, and EPC numbers were determined pre and post treatment. Statin treatment led to a more than twofold increase in DiLDL/lectin-positive cells. In a study in the same edition of the Journal of Clinical investigation, Llevadot et al. [33] used a modified Boyden chamber to assess EPC migration, and used simvastatin, VEGF, or vehicle control as chemoattractants. Simvastatin profoundly enhanced EPC migration (control versus 1 μM simvastatin, $5 \pm 4$ vs. $213 \pm 46$ cells per four high-powered [40×] fields; p <0.01). Migration to 10 μM simvastatin was equivalent to VEGF, and the maximum migration was seen with 1 μM simvastatin. To investigate the effects of simvastatin on EPC mobilization, a chemotactic transwell assay was used. Chemotactic activity was increased by simvastatin, with maximum chemotactic activity observed in the group treated with 1 μM simvastatin (control versus 1 μM simvastatin: $1,137 \pm 148$ vs. $4,681 \pm 598$ cells per 50 μl of lower chamber media; $P < 0.01$). To evaluate EPC mobilization *in vivo*, a murine EPC culture assay was used as a functional index of circulating EPCs. After 4 days in culture, there were significantly more circulating EPCs in the peripheral blood of simvastatin-treated versus control mice ($205 \pm 5$ vs. $147 \pm 7$ cells/mm2; $P < 0.05$) There was no statistically significant difference in the levels of serum cholesterol between treated and control mice. To establish whether these *in vitro* and *in vivo* findings suggesting that the pro-vasculogenic effects of simvastatin on EPCs were associated with augmented neovascularisation, simvastatin therapy was studied in a murine model of corneal injury after bone marrow transplant. Simvastatin treatment resulted in augmented corneal neovascularisation and in more X-gal–positive cells than in the control group. Sections from the simvastatin-treated mice demonstrated more neovascularisation and more extensive incorporation of β-gal–positive cells compared to controls. Quantitative analysis of incorporated β-gal–positive cells revealed that simvastatin enhanced vasculogenesis in neovascular foci of corneas of simvastatin-treated versus control mice ($25.7\% \pm 4.0\%$ versus $7.3\% \pm 2.0\%$ incorporation of β-gal–positive cells; $P < 0.05$). In summary, this study demonstrated that simvastatin enhances EPC migration and chemotaxis *in vitro*, and increases EPC mobilisation from the bone marrow *in vivo*.

A 2011 study by Tousoulis et al. [34] evaluated EPCs (flow cytometry, CD34+/KDR+) in rosuvastatin treatment in patients with heart failure (21 patients assigned rosuvastatin 10mg/day, 18 assigned placebo). Rosuvastatin significantly increased circulating (CD34+/KDR+) EPCs, from 230 (170–380) to 390 (230–520), p = 0.004. This increase in the number of EPCs did not correlate with the decrease in LDL (r=−0.12, p = 0.375), further supporting the theory that the beneficial effect of statins on cardiovascular risk reduction over and above their cholesterol-lowering effect may be mediated by a beneficial effect on EPCs. Subgroup analysis in this study showed that the increase in EPC numbers was similar in subjects with ischaemic and non-ischaemic disease (46% and 36%, respectively, p = 0.63).

A recently published paper examined the effects of rosuvastatin in EPC mediated revascularisation of murine hind limb ischaemia [35]. After treatment of mice with a single low dose of rosuvastatin, circulating EPCs significantly increased from 2 h, peaked at 4 h, declined until 8 h. In a growth-factor reduced Matrigel plug-in assay, rosuvastatin treatment for 5 days induced endothelial lineage differentiation in vivo. Interestingly, the enhanced EPCs and neovascularization stimulated by the statin were blunted in mice that were deficient in endothelial nitric oxide synthase (eNOS). In addition to this, rosuvastatin increased p-Akt/p-eNOS levels in EPCs in vitro, indicating that some of the statin effects on EPCs are dependent on eNOS activity.

# 6. The Effect on EPCs of Other Interventions Known to Reduce Cardiovascular Risk

## 6.1. EPCs and Treatment of Diabetes

Liao et al. [36] quantified EPC number by flow cytometry (CD 45- / CD34+ / VEGFR2+) and endothelial function was assessed by flow-mediated brachial artery dilatation (FMD) in 46 newly diagnosed type 2 diabetic patients and 51 healthy subjects. Metformin was then administered to all patients for 16 weeks.

EPC numbers in the diabetic group were significantly lower than in the control group (p < 0.001), and improved markedly after treatment (p < 0.001). The results of FMD were consistent with EPC variations among the three groups (p < 0.001). In multivariate regression analysis, EPC number was an independent predictor of FMD at baseline (p < 0.05). The absolute changes in EPC number showed a significant correlation with changes in FMD before and after treatment (r = 0.63, p < 0.001). The authors concluded that circulating EPC number correlates well with measures of endothelial function and could be considered as a surrogate biological marker of endothelial function in Type 2 diabetes.

The same group [37] also evaluated EPC response to treatment with gliclazide in type 2 diabetes. EPC number was first quantified in 58 patients with newly diagnosed diabetes and in control subjects, with EPC count lower in the diabetic group than in the control group at baseline (Cell count 1036±94 vs 1624±91, p<0.05). EPC number improved significantly following 12 weeks of gliclazide treatment (Cell count after gliclazide 1411±106 vs 1036±94 at baseline, p<0.05).

## 6.2. EPCs and Treatment of Hypertension

Cacciatore et al. [38] measured EPC number by cell culture and EPC migratory function with a modified Boyden chamber (VEGF as chemoattractant) in hypertensive patients before and after treatment with an ACE-Inhibitor (enalapril 20 mg/day (n=18) or zofenopril 30 mg/day (n=18)). Carotid intimal medial thickness (IMT) was determined by ultrasonography at baseline and after 1 and 5 years of follow-up. EPC number increased during the follow-up (69.6±12.0 vs. 80.6±8.4 for enalapril, 67.5±11.9 vs. 80.9±7.3 for zofenopril, p<0.001), but migrating capacity of EPCs did not change. There was an inverse correlation between circulating EPCs and IMT increase over time. Multiple linear regression model demonstrated that carotid IMT was significantly inversely correlated with EPC (p<0.001) but not with migratory cells after adjusting for confounders.

## 6.3. EPCs and Treatment of Subclinical Hypothyroidism

Subclinical hypothyroidism (SCH) has been associated with an atherogenic lipid profile and endothelial dysfunction (as measured by flow-mediated dilatation), which improves with thyroid hormone replacement therapy [39]. Some studies have shown association of SCH with cardiovascular disease or mortality [40]. Shakoor et al. [41] measured EPCs with cell culture and flow cytometry (CD34+/VEGFR-2+) and EPC function using MMT assay in 20 subjects with subclinical hypothyroidism (median TSH 6.5 mU/L) before and 3 months after thyroxine replacement (median TSH 2.26 mU/L). 20 healthy controls were also studied (HC).

Flow cytometry EPC count was significantly reduced in SCH compared to HC (0.10 vs. 0.39, P < 0.001), whereas after treatment with thyroxine EPC count increased to a similar level as that found in HC (0.26 vs. 0.10, P < 0.001). EPC count as assessed by cell culture showed similar results (HC: 49 [46 –54] SCH at baseline: 47 [36 –54] p=0.06, SCH after treatment: 50 [32–

79] p=0.08) but no difference in EPC function was demonstrated. There was a significant positive correlation between EPCs and free T4 levels (r = 0.38; P = 0.02); high-density lipoprotein cholesterol levels (r = 0.51; P = 0.001); and a negative correlation with TSH concentrations (r = -0.64; P < 0.001). SCH appeared to be the single, most important factor determining lower EPC count when corrected for other cardiovascular risk factors studied, which excludes the possibility that the reduction in EPC count could have been secondary to prevailing risk factors.

# 7. Beneficial Effects of Infusion of EPCs in Animal Models of Ischaemia

Since Asahara's first demonstration of the role of the EPC in an animal model of ischaemia, there have been many subsequent studies confirming the same findings. A further study from the same group [42] transplanted *ex vivo* expanded human DiI-labelled EPCs to athymic nude mice with hind limb ischaemia. Blood flow recovery and capillary density in the ischaemic hind limb were markedly improved, and the rate of limb loss was significantly reduced. Time-course studies demonstrated that peak EPC incorporation was achieved within 3–7 days post administration of EPCs. Review of sections retrieved from the ischaemic limbs animals identified labelled EPCs in up to 56 ± 4.7% (mean ± SEM, range 30–80%) of vessels in a ×10 field. Other than in ischaemic tissue and very rarely in the spleen, EPCs were detected neither in other organs nor the contra lateral non-ischaemic hind limb. This experiment confirmed the concept that *ex vivo* expanded human EPCs may have utility as a "supply-side" strategy for therapeutic neovascularisation.

An experiment from the same laboratory in Boston [43] sought to clarify the extent to which EPCs contribute to adult neovascularisation. The investigators studied the quantitative contribution of EPCs to newly formed vascular structures in an *in vivo* Matrigel plug assay and corneal micropocket assay. They transplanted irradiated mice with bone marrow mononuclear cells from transgenic mice constitutively expressing beta-galactosidase (beta-gal). After 4 weeks a subcutaneous matrigel plug containing fibroblast growth factor 2, or corneal pellet containing VEGF was injected, and mice were sacrificed 7 days later. Bone marrow derived cells in the implants were identified by immunostaining for beta-gal, and 26.5% of all endothelial cells in the new blood vessels in the matrigel plug stained positive. In the corneal implant, 17.7% of the endothelial cells involved in neovascularisation were found to be bone marrow derived. Ki67 staining of the corneal tissue documented that the majority of EPC-derived cells were actively proliferating in situ. These findings suggest that bone marrow derived EPCs make a significant contribution to angiogenic growth factor-induced neovascularisation.

Kawamoto et al. [44] from the same group investigated transplanted human EPCs in an athymic rat model of myocardial ischaemia. Labelled human expanded EPCs were injected intravenously 3 hours after ligation of the left anterior descending artery. After 7 days intravenous lectin was administered and the animals were immediately sacrificed. Fluorescence microscopy revealed that transplanted EPCs accumulated in the ischaemic area and incorporated into foci of myocardial neovascularisation. To determine the impact on left ventricular function, 5 rats (EPC group) were injected intravenously with $10^6$ EPCs 3 hours after ischaemia; 5 other rats (control group) received culture media. Echocardiography showed that the EPC group had ventricular dimensions that were significantly smaller and fractional shortening that was significantly greater in the than in the control group by day 28. Regional wall motion was also better preserved in the EPC group. Histological examination disclosed that capillary density was significantly greater in the EPC group than in the control group. Moreover, the extent of left ventricular scarring was significantly less in rats receiving EPCs than in controls. Immunohistochemistry revealed capillaries that were positive for human-specific endothelial cells. The authors therefore proved that *ex vivo* expanded human EPCs can incorporate into foci of myocardial neovascularisation and have a favourable impact on the preservation of left ventricular function.

Aicher et al. [45] performed a similar experiment with transplanted human EPCs in an athymic rat model of myocardial ischaemia, but labelled the EPCs with a radioactive Indium tracer to monitor tissue distribution of the transplanted EPCs. 8 rats had an induced myocardial infarction, 8 had a sham operation and scintigraphic images were acquired 1, 24, 48, and 96 hours after EPC injection. At 24 to 96 hours after intravenous injection of EPCs, approximately 70% of the radioactivity was localized in the spleen and liver, with only approximately 1% of the radioactivity identified in the heart of sham-operated animals. After myocardial infarction, the heart-to-muscle radioactivity ratio increased significantly, from 1.02+/-0.19 in sham-operated animals to 2.03+/-0.37 after intravenous administration of EPCs. Injection of EPCs into the left ventricular cavity increased this ratio profoundly, from

2.69+/-1.54 in sham-operated animals to 4.70+/-1.55 (P<0.05) in rats with myocardial infarction. On pathological examination, immunostaining of infarcted hearts confirmed that EPCs homed predominantly to the infarct border zone. The authors concluded that, although only a small proportion of radiolabelled EPCs were detected in non-ischaemic myocardium, myocardial infarction profoundly increases homing of transplanted EPCs in vivo.

# 8. Human Studies of EPC Administration

The success of EPC administration for therapeutic benefit in animal studies inevitably led to similar *in vivo* human studies. A particular challenge to the success of such studies, however, is the fact that the patients at greatest need of EPC-based therapies are the very patients with defective EPC number and/or function. Harvesting a healthy population of autologous EPCs for infusion represents a significant challenge as a result. Many researchers have attempted to overcome this difficulty by administering erythropoietin or G-CSF to mobilize EPCs prior to harvesting, or simply harvested bone-marrow cells for administration which, although containing a heterogenous population of cells, do not require culture *in vitro* prior to administration. The Transplantation of Progenitor Cells and Regeneration Enhancement in Acute Myocardial Infarction (TOPCARE-AMI) trial, published in 2002 [46], aimed to evaluate the safety and feasibility of autologous EPC transplantation in humans with ischaemic heart disease. The authors randomly allocated 20 patients with reperfused acute myocardial infarction (AMI) to receive intracoronary infusion of either bone marrow-derived (n=9) or circulating blood-derived progenitor cells (n=11) into the infarct artery 4 days after AMI. EPCs were isolated from the patients' own blood and cultured on fibronectin coated plated prior to re-suspension, and bone marrow derived mononuclear cells (containing heterogeneous cell populations including hematopoietic progenitor cells) were isolated by density gradient centrifugation on the same day as infusion. Cells were administered in a 10ml direct infusion into the infarcted vessel during angiography. Transplantation of progenitor cells was associated with a significant increase in global left ventricular ejection fraction from 51.6 to 60.1% (P=0.003), improved regional wall motion in the infarct zone, and profoundly reduced end-systolic left ventricular volumes at 4-month follow-up. In a non-randomized matched reference group, left ventricular ejection fraction only slightly increased from 51 to 53.5%, and end-systolic volumes remained unchanged. There were no differences for any measured parameter between blood-derived or bone marrow-derived progenitor cells and no signs of an inflammatory response or malignant arrhythmias were observed. Taking into account the significant negatives that there was no randomised control group in this study, and that the study was non-blinded, the study demonstrated that intracoronary infusion of autologous progenitor cells appeared to be feasible and safe, and may beneficially affect the post-infarction remodelling processes. A 5 year follow up of the TOPCARE-AMI study found favourable effects on LV function at follow-up, and further reassurance regarding the safety of intracoronary progenitor cell therapy [47].

In 2003 a group in Japan [48] harvested CD34+ cells from peripheral blood and injected them into patients with critical limb ischaemia that were not suitable for surgical or percutaneous revascularisation. 2 patients were studied, both received granulocyte colony-stimulating factor (G-CSF) prior to the procedure, and cells were injected directly into the muscle of the ischaemic limb. Transcutaneous oxygen pressure in the foot increased following treatment, clinical symptoms improved, and newly-visible collateral blood vessels were directly documented by angiography. Although this study was small, and once again there was no control group, the results added to the hitherto predominantly animal-derived evidence for a key role of the EPC in revascularisation.

In 2005 the first randomized placebo-controlled study of progenitor cells in coronary artery disease was published by Erbs *at al* [49]. The group recruited 26 patients with chronic total occlusion of a coronary artery and evidence of myocardial ischaemia or regional wall motion abnormality on cardiac imaging. GCSF was administered to increase the circulating number of EPCs, and after 4 days mononuclear cells were isolated from peripheral blood by density centrifugation. Cells were cultured for 4 days in endothelial medium in gelatin coated flasks, and 90% of the cells bound lectin and took up DiI-acLDL. After re-canalization of the occlusion, patients were randomly assigned to receive intracoronary infusion of day 4 EPCs or control serum. Coronary flow reserve in response to adenosine was measured in the target vessel at the beginning of the study and after 3 months. Left ventricular function and infarct size were assessed by MRI and metabolism by 18F deoxy-glucose positron emission tomography. In patients administered day 4 EPCs there were statistically significant improvements in coronary flow reserve, and the number of hibernating segments in the target region had declined in the treatment group, whereas no significant changes were observed in the control group. MRI revealed a statistically significant reduction in infarct size by 16% and an increase in LV ejection fraction by 14% in the treatment group, because of an augmented wall motion in the target region. There were similar rates of re-stenosis in both groups (25% vs 27% in control). This study suggested that intracoronary transplantation of EPCs after recanalization results in an improvement of macro- and micro-vascular function and contributes to the recruitment of hibernating myocardium.

Further studies have followed, and in 2012 a Cochrane review of autologous stem/progenitor cell treatment for acute myocardial infarction was performed [50]. The review found that the majority of studies utilized bone-marrow harvested stem cells, and only a minority mobilized progenitor cells into the circulation using a growth factor stimulant (e.g. G-CSF). The review found no mortality or morbidity benefit of stem/progenitor cell treatment over no stem cell treatment, but suggested a moderate improvement in global heart function with stem cell therapy. The authors reported a significant heterogeneity in trial design, and suggested that further larger and more standardized studies are required.

# Conclusion

In summary, EPCs are bone-marrow derived stem cells that are postulated to contribute to post-natal vasculogenesis and to repair of damaged endothelium by incorporation into the vessel wall, secretion of paracrine hormones and stimulation of angiogenesis. Since their original description in 1997, EPCs have been widely studied in cardiovascular disease, and a lower level of circulating EPCs and reduced EPC function in vitro are associated with an increased cardiovascular risk. Medications that reduce an individual's cardiovascular risk have been shown to increase EPC number, and in animal models and human studies EPC infusions have been shown to increase blood flow to areas of ischaemia. The precise future role of EPCs in modern medicine remains a matter of some debate, but the evidence of the important role of EPCs in cardiovascular disease remains incontrovertible. Rather than the direct infusion of EPCs for therapeutic benefit, where other cell lines may prove to be more practical, the future may well rest in the pharmacological manipulation of EPCs, as seen with statin therapy, and the potential utility of EPCs as early and accurate surrogate markers of cardiovascular risk.

# References


[1] Wu KK, Thiagarajan P. Role of endothelium in thrombosis and hemostasis. *Annual review of medicine.* 1996;47:315-31.

[2] Risau W. Mechanisms of angiogenesis. *Nature.* 1997;386(6626):671-4.

[3] Carmeliet P. Angiogenesis in health and disease. Nat Med. 2003;9(6):653-60.

[4] Scott SM, Barth MG, Gaddy LR, Ahl ET, Jr. The role of circulating cells in the healing of vascular prostheses. *J Vasc Surg.* 1994;19(4):585-93.

[5] Wu MH, Shi Q, Wechezak AR, Clowes AW, Gordon IL, Sauvage LR. Definitive proof of endothelialization of a Dacron arterial prosthesis in a human being. *J Vasc Surg.* 1995;21(5):862-7.

[6] Rafii S, Oz MC, Seldomridge JA, Ferris B, Asch AS, Nachman RL, et al. Characterization of hematopoietic cells arising on the textured surface of left ventricular assist devices. *The Annals of Thoracic Surgery.* 1995;60(6):1627-32.

[7] Asahara T, Murohara T, Sullivan A, Silver M, van der Zee R, Li T, et al. Isolation of Putative Progenitor Endothelial Cells for Angiogenesis. *Science.* 1997;275(5302):964-6.

[8] Hill JM, Zalos G, Halcox JP, Schenke WH, Waclawiw MA, Quyyumi AA, et al. Circulating endothelial progenitor cells, vascular function, and cardiovascular risk. *N Engl J Med.* 2003;348(7):593-600.

[9] Khan SS, Solomon MA, McCoy JP, Jr. Detection of circulating endothelial cells and endothelial progenitor cells by flow cytometry. *Cytometry.* 2005;64(1):1-8.

[10] Fadini G, Sartore S, Schiavon M, Albiero M, Baesso I, Cabrelle A, et al. Diabetes impairs progenitor cell mobilisation after hindlimb ischaemia–reperfusion injury in rats. *Diabetologia.* 2006;49(12):3075-84.

[11] Fadini GP, Miorin M, Facco M, Bonamico S, Baesso I, Grego F, et al. Circulating Endothelial Progenitor Cells Are Reduced in Peripheral Vascular Complications of Type 2 Diabetes Mellitus. *Journal of the American College of Cardiology.* 2005;45(9):1449-57.

[12] Fadini GP, Sartore S, Baesso I, Lenzi M, Agostini C, Tiengo A, et al. Endothelial Progenitor Cells and the Diabetic Paradox. *Diabetes Care.* 2006;29(3):714-6.

[13] Schmidt-Lucke C, Rossig L, Fichtlscherer S, Vasa M, Britten M, Kamper U, et al. Reduced Number of Circulating Endothelial Progenitor Cells Predicts Future Cardiovascular Events: Proof of Concept for the

Clinical Importance of Endogenous Vascular Repair. *Circulation.* 2005;111(22):2981-7.

[14]  Eizawa T, Murakami Y, Matsui K, Takahashi M, Muroi K, Amemiya M, et al. Circulating endothelial progenitor cells are reduced in hemodialysis patients. *Curr Med Res Opin.* 2003;19(7):627-33.

[15]  Vasa M, Fichtlscherer S, Aicher A, Adler K, Urbich C, Martin H, et al. Number and Migratory Activity of Circulating Endothelial Progenitor Cells Inversely Correlate With Risk Factors for Coronary Artery Disease. *Circ Res.* 2001;89(1):1e-7.

[16]  Van Craenenbroeck EM, Conraads VM, Van Bockstaele DR, Haine SE, Vermeulen K, Van Tendeloo VF, et al. Quantification of circulating endothelial progenitor cells: A methodological comparison of six flow cytometric approaches. *Journal of immunological methods.* 2008;332(1-2):31-40.

[17]  Jialal I, Devaraj S, Singh U, Huet BA. Decreased number and impaired functionality of endothelial progenitor cells in subjects with metabolic syndrome: implications for increased cardiovascular risk. *Atherosclerosis.* 2010;211(1):297-302.

[18]  Loomans CJ, de Koning EJ, Staal FJ, Rookmaaker MB, Verseyden C, de Boer HC, et al. Endothelial progenitor cell dysfunction: a novel concept in the pathogenesis of vascular complications of type 1 diabetes. *Diabetes.* 2004;53(1):195-9.

[19]  Vaughan EE, Liew A, Mashayekhi K, Dockery P, McDermott J, Kealy B, et al. Pretreatment of endothelial progenitor cells with osteopontin enhances cell therapy for peripheral vascular disease. *Cell Transplant.* 2012;21(6):1095-107.

[20]  Tepper O, Galiano R, Capla J, Kalka C, Gagne P, Jacobowitz G, et al. Human endothelial progenitor cells from type II diabetics exhibit impaired proliferation, adhesion, and incorporation into vascular structures. *Circulation.* 2002;106(22):2781-6.

[21]  King TF, Bergin DA, Kent EM, Manning F, Reeves EP, Dicker P, et al. Endothelial progenitor cells in mothers of low-birthweight infants: a link between defective placental vascularization and increased cardiovascular risk? *J Clin Endocrinol Metab.* 2013;98(1):E33-9.

[22]  Vasa M, Fichtlscherer S, Aicher A, Adler K, Urbich C, Martin H, et al. Number and migratory activity of circulating endothelial progenitor cells inversely correlate with risk factors for coronary artery disease. *Circ Res.* 2001;89(1):E1-7.

[23] Yue WS, Wang M, Yan GH, Yiu KH, Yin L, Lee SW, et al. Smoking is associated with depletion of circulating endothelial progenitor cells and elevated pulmonary artery systolic pressure in patients with coronary artery disease. *Am J Cardiol.* 2010;106(9):1248-54.

[24] Schmidt-Lucke C, Rossig L, Fichtlscherer S, Vasa M, Britten M, Kamper U, et al. Reduced number of circulating endothelial progenitor cells predicts future cardiovascular events: proof of concept for the clinical importance of endogenous vascular repair. *Circulation.* 2005;111(22):2981-7.

[25] Werner N, Kosiol S, Schiegl T, Ahlers P, Walenta K, Link A, et al. Circulating endothelial progenitor cells and cardiovascular outcomes. *N Engl J Med.* 2005;353(10):999-1007.

[26] Chen M, Sheu J, Wang P, Chen C, Kuo M, Hsieh C, et al. Complications impaired endothelial progenitor cell function in Type 2 diabetic patients with or without critical leg ischaemia: implication for impaired neovascularization in diabetes. *Diabet Med.* 2009;26(2): 134-41.

[27] Huang C, Zhang L, Wang Z, Pan H, Zhu J. Endothelial progenitor cells are associated with plasma homocysteine in coronary artery disease. *Acta Cardiol.* 2011;66(6):773-7.

[28] Schlager O, Giurgea A, Schuhfried O, Seidinger D, Hammer A, Groger M, et al. Exercise training increases endothelial progenitor cells and decreases asymmetric dimethylarginine in peripheral arterial disease: A randomized controlled trial. *Atherosclerosis.* 2011.

[29] Heida NM, Muller JP, Cheng IF, Leifheit-Nestler M, Faustin V, Riggert J, et al. Effects of obesity and weight loss on the functional properties of early outgrowth endothelial progenitor cells. *J Am Coll Cardiol.* 2010;55(4):357-67.

[30] Wu VC, Lo SC, Chen YL, Huang PH, Tsai CT, Liang CJ, et al. Endothelial progenitor cells in primary aldosteronism: a biomarker of severity for aldosterone vasculopathy and prognosis. *J Clin Endocrinol Metab.* 2011;96(10):3175-83.

[31] Quist-Paulsen P. Statins and inflammation: an update. *Curr Opin Cardiol.* 2010;25(4):399-405.

[32] Dimmeler S, Aicher A, Vasa M, Mildner-Rihm C, Adler K, Tiemann M, et al. HMG-CoA reductase inhibitors (statins) increase endothelial progenitor cells via the PI 3-kinase/Akt pathway. *J Clin Invest.* 2001;108(3):391-7.

[33] Llevadot J, Murasawa S, Kureishi Y, Uchida S, Masuda H, Kawamoto A, et al. HMG-CoA reductase inhibitor mobilizes bone marrow--derived endothelial progenitor cells. *J Clin Invest.* 2001;108(3):399-405.

[34] Tousoulis D, Andreou I, Tsiatas M, Miliou A, Tentolouris C, Siasos G, et al. Effects of rosuvastatin and allopurinol on circulating endothelial progenitor cells in patients with congestive heart failure: the impact of inflammatory process and oxidative stress. *Atherosclerosis.* 2011;214(1):151-7.

[35] Zhou J, Cheng M, Liao YH, Hu Y, Wu M, Wang Q, et al. Rosuvastatin Enhances Angiogenesis via eNOS-Dependent Mobilization of Endothelial Progenitor Cells. *PLoS One.* 2013;8(5):e63126.

[36] Liao YF, Chen LL, Zeng TS, Li YM, Fan Y, Hu LJ, et al. Number of circulating endothelial progenitor cells as a marker of vascular endothelial function for type 2 diabetes. *Vasc Med.* 2010;15(4):279-85.

[37] Chen LL, Yu F, Zeng TS, Liao YF, Li YM, Ding HC. Effects of gliclazide on endothelial function in patients with newly diagnosed type 2 diabetes. *Eur J Pharmacol.* 2011;659(2-3):296-301.

[38] Cacciatore F, Bruzzese G, Vitale DF, Liguori A, de Nigris F, Fiorito C, et al. Effects of ACE inhibition on circulating endothelial progenitor cells, vascular damage, and oxidative stress in hypertensive patients. *Eur J Clin Pharmacol.* 2011.

[39] Razvi S, Ingoe L, Keeka G, Oates C, McMillan C, Weaver JU. The beneficial effect of L-thyroxine on cardiovascular risk factors, endothelial function, and quality of life in subclinical hypothyroidism: randomized, crossover trial. *J Clin Endocrinol Metab.* 2007;92(5): 1715-23.

[40] Hak AE, Pols HA, Visser TJ, Drexhage HA, Hofman A, Witteman JC. Subclinical hypothyroidism is an independent risk factor for atherosclerosis and myocardial infarction in elderly women: the Rotterdam Study. *Ann Intern Med.* 2000;132(4):270-8.

[41] Shakoor SK, Aldibbiat A, Ingoe LE, Campbell SC, Sibal L, Shaw J, et al. Endothelial progenitor cells in subclinical hypothyroidism: the effect of thyroid hormone replacement therapy. *J Clin Endocrinol Metab.* 2010;95(1):319-22.

[42] Kalka C, Masuda H, Takahashi T, Kalka-Moll WM, Silver M, Kearney M, et al. Transplantation of ex vivo expanded endothelial progenitor cells for therapeutic neovascularization. *Proc Natl Acad Sci U S A.* 2000;97(7):3422-7.

[43] Murayama T, Tepper OM, Silver M, Ma H, Losordo DW, Isner JM, et al. Determination of bone marrow-derived endothelial progenitor cell significance in angiogenic growth factor-induced neovascularization in vivo. *Exp Hematol.* 2002;30(8):967-72.

[44] Kawamoto A, Gwon HC, Iwaguro H, Yamaguchi JI, Uchida S, Masuda H, et al. Therapeutic potential of ex vivo expanded endothelial progenitor cells for myocardial ischemia. *Circulation.* 2001;103(5): 634-7.

[45] Aicher A, Brenner W, Zuhayra M, Badorff C, Massoudi S, Assmus B, et al. Assessment of the tissue distribution of transplanted human endothelial progenitor cells by radioactive labeling. *Circulation.* 2003;107(16):2134-9.

[46] Assmus B, Schachinger V, Teupe C, Britten M, Lehmann R, Dobert N, et al. Transplantation of Progenitor Cells and Regeneration Enhancement in Acute Myocardial Infarction (TOPCARE-AMI). *Circulation.* 2002;106(24):3009-17.

[47] Leistner DM, Fischer-Rasokat U, Honold J, Seeger FH, Schächinger V, Lehmann R, et al. Transplantation of progenitor cells and regeneration enhancement in acute myocardial infarction (TOPCARE-AMI): final 5-year results suggest long-term safety and efficacy. *Clin Res Cardiol.* 2011;100(10):925-34.

[48] Kudo FA, Nishibe T, Nishibe M, Yasuda K. Autologous transplantation of peripheral blood endothelial progenitor cells (CD34+) for therapeutic angiogenesis in patients with critical limb ischemia. *Int Angiol.* 2003;22(4):344-8.

[49] Erbs S, Linke A, Adams V, Lenk K, Thiele H, Diederich KW, et al. Transplantation of blood-derived progenitor cells after recanalization of chronic coronary artery occlusion: first randomized and placebo-controlled study. *Circ Res.* 2005;97(8):756-62.

[50] Clifford DM, Fisher SA, Brunskill SJ, Doree C, Mathur A, Watt S, et al. Stem cell treatment for acute myocardial infarction. *Cochrane Database Syst Rev.* 2012;2:CD006536.

In: Progenitor Cells
Editors: P. M. Horton, B. E. Lawrence

ISBN: 978-1-62808-994-3
© 2013 Nova Science Publishers, Inc.

*Chapter 2*

# Characterization, Isolation, Expansion and Clinical Therapy of Human Corneal Epithelial Stem/Progenitor Cells

*De-Quan Li*[1,*], *Zhichong Wang*[2],
*Kyung-Chul Yoon*[1,3] *and Fang Bian*[1,4]

[1]Ocular Surface Center, Cullen Eye Institute,
Department of Ophthalmology and Center for Stem Cell
and Regenerative Medicine, Center for Cell and Gene Therapy,
Baylor College of Medicine, Houston, TX, US
[2]Zhongshan Ophthalmic Center, State Key laboratory of
Ophthalmology, Sun Yat-Sen University, Guangzhou, China
[3]Department of Ophthalmology,
Chonnam National University Medical School and Hospital,
Gwangju, South Korea
[4]Department of Ophthalmology,
Union Hospital of Tongji Medical College,
Huazhong Science and Technology University, Wuhan, China

---

[*] Correspondence to: De-Quan Li, M.D., Ph.D., Associate Professor, Department of Ophthalmology, Baylor College of Medicine, 6565 Fannin Street, NC-205, Houston, TX 77030, USA. Tel: (713) 798-1123, Fax: (713) 798-1457, E-mail: dequanl@bcm.tmc.edu.

# Abstract

Stem cells can be defined as cells that have the capacity to self-renew and the ability to generate differentiated progeny or multiple cell lineages. True stem cells can turn into any type of cells, while progenitor cells are more or less committed to becoming cell types of a particular tissue. Human corneal epithelial stem cells (CESCs) represent a great example and model of adult stem or progenitor cells. Human CESCs have been identified to locate in the basal epithelial layer of the limbus, and thus also referred as to limbal stem cells. We would like to use the both terms, stem and progenitor cells in this chapter based on previous use in the literature for more than two decades. Although the CESCs have been identified to reside at the limbus and many stem cell markers have been proposed, there is no consensus to date regarding the definitive markers for CESCs, and identification and isolation of these cells are still challenging. Based on evaluation of a variety of proposed markers, we have characterized that the CESCs located in the basal layer of human limbal epithelium are small primitive cells expressing three patterns of molecular markers, which represent a unique phenotype of putative corneal epithelial stem or progenitor cells.

Based on adult stem cell criteria and the putative limbal stem cell phenotype, our group has attempted to enrich for human CESCs through novel approaches including cell-sizing, adhering to extracellular matrix collagen type IV, and cell sorting for side population or for expression of ABCG2 or connexin 43 cell surface markers. The 5 clonogenic populations isolated from limbal epithelium and its cultures by different methods show the properties that are characteristics of adult stem/progenitor cells: 1) relatively undifferentiated, 2) high proliferative potential, 3) self-renewal. Expansion and cultivation of corneal epithelial progenitor cells have been achieved using different methods, such as limbal tissue explant culture, and limbal epithelial cell suspension co-culture with mouse 3T3 fibroblast feed layer. To avoid the use of xeno-components, two cell lines of commercial human fibroblasts have been identified that support human corneal epithelial regeneration, and have potential use in replacing mouse 3T3 cells for corneal tissue bioengineering.

The concept of CESCs has formed the basis for identifying a class of blinding diseases that display features of corneal epithelial stem cell deficiency or limbal stem cell deficiency (LSCD), where the limbal epithelium is damaged. LSCD is characterized by persistent or recurrent epithelial defects, ulceration, corneal vascularization, chronic inflammation, scarring, and conjunctivalization (conjunctival epithelial ingrowth). Only transplantation of CESCs can restore vision. Due to an increasing shortage of corneal donors, corneal tissue engineering is becoming an important discipline that holds great promise for corneal reconstruction. CESCs and optical substrates are known to be the most important factors for corneal tissue bioengineering in regenerative medicine. Our team has recently explored the utilization of natural donor corneal stroma in corneal tissue engineering. In combination with fresh limbal epithelium containing stem cells, and the donor corneal stroma, a great source of natural optical substrate, we developed a native-like corneal equivalent construct with proliferative potential. This corneal construct provides a new clinical cell therapy for corneal reconstruction.

**Keywords:** Adult stem cell, progenitor cell, progenitor markers, corneal epithelium, progenitor cell therapy

# Concept of Tissue Specific Adult Stem Cells and Progenitor Cells

Stem cells can be defined as cells that have the capacity to self-renew and the ability to generate differentiated progeny or multiple cell lineages. While recent studies indicate that embryonic stem cells have pluripotent potential, it is also recognized that there are organ-specific stem cells residing in many adult tissues, including the bone marrow, liver, brain, intestine, skin and cornea [1-3]. True stem cells can turn into any type of cells, while progenitor cells are more or less committed to becoming cell types of a particular tissue. In general, embryonic stem cells are true stem cells that can turn into any type of cell, while some adult stem cells may actually represent progenitor cells in that they may turn into cells of the specific tissue in which they reside, but not into any other cell types. Adult stem cells, somatic stem cells, or organ-specific stem cells, are small subpopulations with a number of characteristics in morphology, phenotype and growth potential. Widely accepted criteria for defining adult stem cells include (1) quiescent slow cycling or long cell cycle time during homeostasis in vivo; (2) undifferentiated small cells with primitive cytoplasm; (3) high proliferative and/or pluripotent potential after wounding or placement in culture; (4) the capacity for self-renewal through the life span; and (5) the ability to regenerate functional tissues [1-5].

Adult stem cells may originate differentiated progeny through asymmetric divisions. The asymmetric cell division gives rise to one daughter stem cell (for self-renewal) and one non-stem cell already committed to differentiation into specific phenotypes (for tissue regeneration). Hence, not all progenitor cells are adult stem cells. Indeed, there are adult stem cells and "transient-amplifying cells" or committed precursor cells or committed progenitor cells. The latter, with more rapid though limited proliferation, low self-renewal and restricted differentiation, increase the number of differentiated cells produced by one adult stem cells division. Therefore, the transient-amplifying cells, also referred to as transit-amplifying cells or transit cells, are committed progenitors between the adult stem cells and all their terminally differentiated cells [2, 6].

The adult stem cells reside in a specialized physical location known as a niche [1, 6, 7], which constitutes a three-dimensional microenvironment containing the stem cells, neighboring differentiated cell types, extracellular matrix and mesenchymal cells. Niche cells provide a sheltering environment that sequesters stem cells from differentiation stimuli, apoptotic stimuli, and other stimuli that would challenge stem cell reserves. The niche also safeguards against excessive stem cell production that could lead to cancer. Thus the niche is a specific location in a tissue where stem cells can reside for an indefinite period of time, produce progeny cells while self-renewing, and be able to substitute damaged cells and to intervene in maintaining the structural and functional integrity of the tissues.

# Human Corneal Epithelial Stem/Progenitor Cells

Human corneal epithelial stem cells (CESCs) represent a great example and model of adult stem or progenitor cells. Human CESCs have been identified to locate in the basal epithelial layer of limbus, a 1.5-2 mm wide transitional zone that straddles the cornea and bulbar conjunctiva, and thus also referred as to limbal stem cells [4, 8-10]. We would like to use the both terms, stem and progenitor cells in this chapter based on previous use in the literature for more than two decades.

The corneal epithelium includes, from its superficial aspect, approximately one to three layers of flattened cells called squames, two to three layers of suprabasal or wing cells, and a single layer of columnar basal cells. Although it was observed in 1971 that corneal epithelial cells are repopulated by centripetal migration of cells from the limbus [11], it was not until 1986 that Schermer and colleagues [12] proposed that stem cells localized in the limbus were responsible for generating and maintaining the corneal epithelium.

The supporting data for the limbal location of CESCs include: (1) the limbal basal epithelium contains quiescent slow-cycling cells identified as the "label-retaining cells" following pulse-chase labeling of all cells with a DNA precursor, such as [3H]-thymidine or bromodeoxyuridine (BrdU) [13]; (2) the limbal basal cells lack the corneal epithelial differentiation-associated cytokeratin (K) pair K3 [12] and K12 [14]; (3) the limbal basal epithelium exhibits much higher proliferative potential in culture than does the central corneal epithelium [13, 15, 16]; (4) abnormal corneal epithelial wound healing occurs with conjunctivalization, vascularization and chronic inflammation when the limbal epithelium is partially [17] or completely defective [18] in human patients and experimental models; (5) limbal cells are essential for the long-term maintenance of the central corneal epithelium and they can be used to reconstitute the entire corneal epithelium in patients with limbal stem cell deficiencies [9, 19]. Collectively, these data leave little doubt that CESCs reside in the limbus and exhibit the full complement of well-defined keratinocyte stem cell properties. Thus, CESCs are also referred to as limbal stem cells based on their location.

# The Phenotype of Corneal Epithelial Progenitor Cells

The limbal basal epithelium consists of at least three functionally different cell types: stem cells, transient amplifying cells that are an intermediate population of progenitor cells, and terminally differentiated cells. The stem cells are only a small subpopulation, estimated as less as <1% of these limbal basal cells [20, 21], perhaps as few as 100 cells/limbus are true stem cells [10, 22]. Although CESCs have been identified to reside at the limbus for more than two decades [12] and many stem cell markers have been proposed, there is still no consensus regarding the definitive markers for limbal stem cells [10, 20, 21, 23], and identification and isolation of these cells are challenging. Based on evaluation of a variety of proposed markers, it has been characterized that basal cells at human limbal epithelium are small primitive cells expressing three patterns of molecular markers [24]: (1) exclusively positive for ABCG2, p63, N-cadherin, integrin α9, NGF, TrkA, GDNF and GFRα-1 by a subset of basal cells; (2) relatively higher expression of integrin β1, EGFR, K19 and α-enolase by most basal cells, and (3) lack of expression of nestin, E-cadherin, connexin 43, involucrin, K3 and K12. These three patterns of basal cell markers represent a unique phenotype for identification of putative limbal stem cells [24, 25].

ATP-binding cassette, sub-family G (WHITE), member 2 (ABCG2), a member of the ATP binding cassette transporters, formally known as breast cancer resistance protein 1 (BCRP1), has been identified as a molecular determinant for the side population that is enriched in hematopoeitic stem cells, and has been proposed as a universal marker for stem cells [26] including corneal epithelial cells [20, 27]. The nuclear transcription factor p63, a member of the p53 family, is highly expressed in the basal cells of many human epithelial tissues and the truncated dominant-negative △Np63 isoform is the predominant species in these cells [28]. It was reported that p63 knockout mice lack stratified epithelia and contain clusters of terminally differentiated keratinocytes on the exposed dermis [28], and that p63 expression is associated with proliferative potential in human keratinocytes [29, 30].

N-cadherin is a member of the classical cadherin family and has been previously demonstrated to be expressed by hematopoietic stem cells. Recently, N-cadherin has been demonstrated to be exclusively expressed by partial limbal basal cells [25]. Although neurotrophic factors are defined as polypeptides that maintain neuronal cells, they possess a range of functions in survival and self renewal of stem cells outside the nervous system [31, 32]. Our group has also identified that the nerve growth factor (NGF) and glial cell-derived neurotrophic factor (GDNF), and their corresponding receptors, TrkA and GFRα-1, were exclusive localized to a subpopulation of basal limbal epithelial cells where stem cells reside [33-36].

Integrins are cell surface adhesion molecules, that are composed of noncovalently associated heterodimer transmembrane receptors consisting of an α and a β subunit that bind to extracellular matrix proteins. Cell-matrix adhesion is largely mediated by integrins, which are the major components of stable adhesions to the basement membrane in hemiadherent junctions. Integrin α9 was reported to localize to the basal cells of the epidermis, conjunctiva and corneal limbus after birth and into adulthood in the developing ocular surface of mice [37]. Integrin β1 was previously proposed as a putative stem cell marker for epidermal keratinocytes. Integrin β1 enriched human epidermal basal cells from both keratinocyte culture and foreskin biopsies were demonstrated to have a higher colony-forming efficiency than unfractionated cells [38, 39]. In addition, the slow-cycling

BrdU label retaining cells in the adult mouse cornea are enriched in cells that express high levels of integrin $\beta 1$ and $\beta 4$ and little integrin $\alpha 9$ [40]. It has been reported that limbal basal epithelial cells express higher levels of EGFR than the limbal suprabasal cells, and the more mature and differentiated cells express the lowest levels of EGFR [41, 42]. As a member of the cytokeratin family of intermediate filaments, K19 has been suggested as a marker for the epidermal stem cells in skin hair follicles. K19 was expressed in the hair follicle and was absent from the interfollicular epidermis at hairy sites. K19 was also noted in the slow cycling [3H]-thymidine-label-retaining cells by double-labeling experiments [43]. A cytoplasmic glycolytic enzyme, $\alpha$-enolase, was proposed as a candidate of CESC marker, which was localized to the limbal basal cells, as well as the basal cells of other stratified epithelia [10, 44].

Cell to cell communication plays an important role in cellular development and differentiation. Gap junctional communication is mediated by a family of related amphipathic polypeptides called connexins. The presence of these intercellular communicating channels allows direct passive diffusion of low molecular weight solutes between neighboring cells [45]. E-cadherin is a transmembrane $Ca^{2+}$-dependent homophilic adhesion receptor that plays an important role in cell-cell adhesion. E-cadherin mediated cell-cell contact results in cell activation and an increase in key signaling molecules involved in cell differentiation and survival [46]. These findings suggest that connexin 43 and E-cadherin are expressed by differentiated epithelial cells, and the absence of these intercellular communication molecules in limbal basal cells may be a inherent feature of stem cells, reflecting the need for stem cell to maintain the uniqueness of its own intracellular environment [45, 46]. K3 and K12 are well known as corneal specific markers [12, 14, 47]. Involucrin, a structural component of the cornified cell envelope proteins [48], is another differentiation associated marker [49].

These three patterns of basal cell markers represent a unique phenotype for identification of putative limbal stem cells [24, 25]. The presence of p63 in limbal basal cells appears to represent a higher proliferative potential, the presence of ABCG2 may be a feature of limbal stem cells that protects them from damage by drugs and toxins, and the NGF and GDNF with their receptor co-localized in the limbal basal epithelium may serve as critical survival factors for the stem cells. The higher expression of integrin $\alpha 9$ and $\beta 1$ may indicate the strong adhesion of limbal basal cells to their underlying extracellular matrix and may explain the resistance of limbus to shear forces. The presence of high levels of EGFR might allow these basal cells to be rapidly stimulated by growth factors to undergo cell division during development and following wounding. With more abundant integrins ($\beta1$ and $\alpha9$) and lack of connexin 43 and E-cadherin, the limbal stem cells in vivo are likely to strongly adhere to the extracellular matrix, perhaps keeping them in their niche, and yet they are less adhesive to one another, enabling individual stem cells to be rapidly mobilized and exit from their niche for self renewal. Furthermore, the lack of expression of K3, K12 and involucrin indicates that the limbal basal cells are relatively undifferentiated.

## Isolation and Expansion of Corneal Epithelial Progenitor Cells

Although the concept that CESCs reside in the limbus has been recognized for two decades, isolation of limbal stem cells has not been accomplished due to the lack of a truly unique stem cell marker. Based on adult stem cell criteria and the putative limbal stem cell phenotype, our group has attempted to enrich for human limbal stem cells through novel approaches including cell-sizing, adhering to extracellular matrix collagen type IV, and cell sorting for side population or for expression of ABCG2 or connexin 43 cell surface markers. As summarized in Table 1, these studies have successfully isolated 5 clonogenic cell populations from limbal epithelium and/or its cultures [27, 50-52], which show properties that are characteristics of adult stem/progenitor cells: (1) relatively undifferentiated: they expressed higher levels of stem cell associated markers (ABCG2, p63, or integrin $\beta1$) and negative levels of differentiation markers (K3, K12, involucrin or connexin 43) at both protein and mRNA levels; (2) high proliferative potential: they showed greater clonal forming efficiency (2-4 fold) and growth capacity in culture; and (3) self-renewal: they also contained quiescent slow-cycling BrdU label-retaining cells in culture, an intrinsic characteristic of stem cells, which was enriched 3-5 fold from unfractionated whole cells.

Ex vivo expansion of adult stem/progenitor cells from small tissue biopsies is an important approach to cultivate and preserve stem cells for stem/progenitor cell therapy and tissue engineering. There are two popular culture systems, explant culture and single cell culture, currently used for human corneal epithelial cultivation [53, 54].

**Table 1. Properties of isolated clonogenic populations enriched in corneal epithelial stem cells**

| Properties | Small cells | RAC | SP | ABCG2+ | Cx43dim | ALL |
|---|---|---|---|---|---|---|
| Year of Publication | 2006 [50] | 2005 [51] | 2005 [27] | 2005 [27] | 2006 [52] | |
| Markers | Size | Adhesion | Hoechst | ABCG2 | Connexin 43 | None |
| Population (%) | $11.0 \pm 4.5$ | $10.4 \pm 1.5$ | $2.90 \pm 1.14$ | $2.31 \pm 1.83$ | $9.3 \pm 1.7$ | 100 |
| BrdU-LRC (%) | $11.6 \pm 1.5$ | $16.0 \pm 2.5$ | | | $13.49 \pm 3.55$ | $3.2 \pm 0.6$ |
| Positive Cells (%) | | | | | | |
| ABCG2 + | $38.12 \pm 1.63$ | | | | $26.7 \pm 10.5$ | $10.2 \pm 5.5$ |
| p63 + | $37.26 \pm 5.22$ | $47.5 \pm 8.6$ | | | $63.3 \pm 13.7$ | $19.6 \pm 4.5$ |
| Integrin $\beta1$ + | | $57.1 \pm 9.8$ | | | $85.4 \pm 15.2$ | $26.6 \pm 3.1$ |
| Connexin 43 + | | | | | $5.1 \pm 3.0$ | $59.3 \pm 15.2$ |
| Involucrin + | $1.86 \pm 0.2$ | $1.8 \pm 0.6$ | | | $15.2 \pm 5.9$ | $41.8 \pm 5.6$ |
| Keratin 3 + | $5.21 \pm 0.50$ | | | | $12.5 \pm 5.3$ | $40.9 \pm 5.0$ |
| mRNA Levels | | | | | | |
| ABCG2 | +++ | +++ | ++ | ++ | +++ | + |
| $\Delta$Np63 | +++ | +++ | +++ | +++ | +++ | ++ |
| Integrin $\beta1$ | | ++++ | | | +++ | ++ |
| Connexin 43 | | | | | + | ++ |
| Involucrin | + | - | | | + | ++ |
| Keratin 3 | + | | | | - | ++ |
| Keratin 12 | + | - | | | | ++ |
| GAPDH | ++++ | ++++ | ++++ | ++++ | ++++ | ++++ |
| CFE on Day 6 (%) | $5.43 \pm 0.98$ | $3.8 \pm 0.42$ | $4.84 \pm 1.56$ | $5.67 \pm 0.54$ | $8.7 \pm 1.15$ | $2.0 \pm 0.51$ |
| Confluent Days | 12-14 | 10-14 | 14 | 14 | 12-14 | 18-24 |

RAC: rapidly adherent cells; SP: side population; ALL: unfractioned whole limbal epithelial cells served as controls.

We have evaluated the growth potential, morphology and molecular phenotypes of corneal epithelial cells *ex vivo* expanded from these two culture systems, limbal explant culture or limbal single cell suspension culture on a mitomycin C treated mouse 3T3 fibroblast feeder layer. We observed that the phenotypes of corneal epithelial cells, ranged from basal cells to superficial differentiated cells, are well maintained in both culture systems, and the slow-cycling BrdU-label retaining cells, which are characteristics of stem cells, were identified in the cultures [55].

This 3T3 fibroblast co-culture system has proven to be the most powerful system for human epithelial cultivation [56-59], including human ocular surface epithelia [60-63], over the past 3 decades. Based on our observation, the proliferative rate and regenerative capacity of human corneal epithelial cells on the 3T3 fibroblast feeder layer are 10-50 times higher than explant cultures [55]. As few as 200-1000 limbal epithelial cells, which can be obtained from as small as 0.5 mm$^2$ of limbal biopsy, can regenerate a whole human corneal epithelium in 2 weeks with this co-culture system. Human corneal epithelium regenerated on 3T3 fibroblast feeder layer preserved more stem/progenitor cells than explant cultures [54, 55, 62].

However, human epithelia generated on mouse 3T3 fibroblasts may bear a risk of transmitting pathogens and as a consequence their use is not permitted for cultivating corneal epithelia for human transplantation. To develop this powerful co-culture system for clinical use without xeno-components, mouse 3T3 fibroblasts must be replaced by human fibroblasts. We have evaluated nine human fibroblasts from different source, including 3 primary cultured ocular surface fibroblasts and 6 human fibroblast cell lines, and successfully identified two commercial cell lines of human newborn foreskin fibroblasts, Hs68 and CCD1112Sk, which showed comparative efficiency to a 3T3 feeder layer in supporting human corneal epithelial cell growth and regeneration.

Based on evaluation of epithelial clonal colony forming efficiency and epithelial regeneration capacity, human newborn foreskin fibroblasts, Hs68 and CCD1112Sk, showed strong supportive capacity, similar to the mouse 3T3 feeder layer (Figure 1). The seeded epithelial cells (1x10$^4$ cells) grew rapidly from colonies to confluence reaching about 1.87-2.41x10$^6$ cells in 12-14 days, which represented 187-241 fold expansion with over 7-8 doublings. Cell morphology and immunofluorescent staining further revealed that the regenerated epithelium on these human feeders resembled the phenotype of human corneal epithelium containing undifferentiated progenitor cells. The regenerated epithelium expressed three progenitor cell markers, p63, integrin β1 and EGFR, in addition to corneal epithelial specific and differentiated markers K3, K14 and Cx43. The cultivated corneal epithelium on human fibroblast feeder layers resembles the phenotypes of human corneal epithelium containing progenitor cells. Human feeders have potential in use for tissue bioengineering, not only for corneal epithelium, but also possibly for other epithelial tissues.

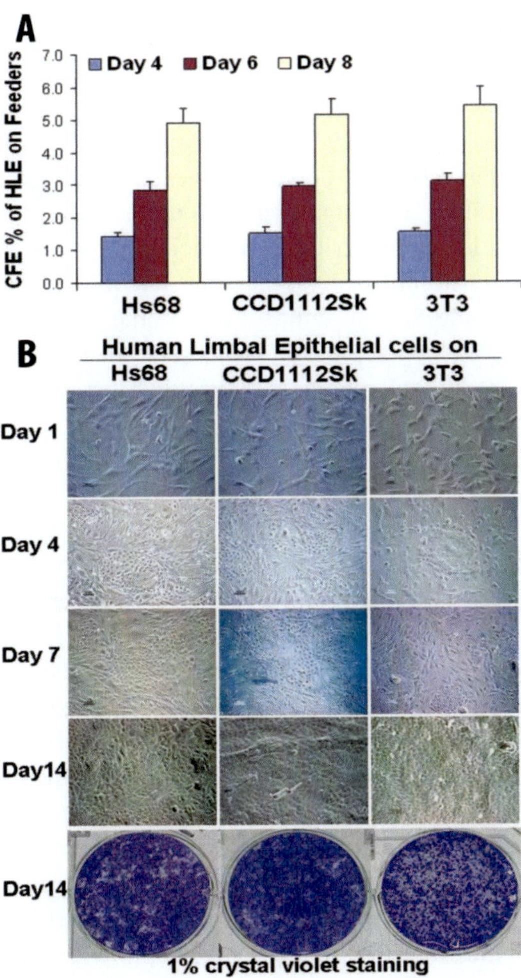


Figure 1. Human limbal epithelial cells (HLE) grown on feeder layers of Hs68, CCD1112Sk or 3T3 fibroblasts after seeded at 1x103 cells/cm2 in 6-well plate. A. Colony forming efficiency (CFE) of HLE on three fibroblasts in days 4-8. B. Phase images showing a comparative efficiency of Hs68, CCD1112Sk and 3T3 fibroblasts as a feeder layer in supporting clonogenicity and growth capacity of HLE on days 1-14.

# Corneal Epithelial Stem Cell Deficiency

The concept of CESCs has formed the basis for identifying a class of corneal blinding diseases that display features of corneal epithelial stem cell deficiency or dysfunction, which is often referred as to limbal stem cell deficiency (LSCD). These conditions, in which the limbal basal epithelium and/or limbal stroma are damaged, display common features of defective stem cells with conjunctivalization of the cornea and recurrent epithelial breakdown. The etiology of LSCD is diverse and can be a variety of hereditary or acquired disorders [18, 19, 64, 65]. Inherited disorders include aniridia keratitis [66], familial dysautonomia, ectodermal dysplasia and keratitis associated with multiple endocrine deficiencies, in which limbal stem cells may be congenitally absent or dysfunctional. Acquired conditions that may result in LSCD include Stevens-Johnson syndrome, chemical injuries, ocular cicatricial pemphigoid, contact lens-induced keratopathy, multiple surgeries or cryotherapies to the limbal region, neurotrophic keratopathy and peripheral ulcerative keratitis [9, 67].

LSCD is characterized by persistent or recurrent epithelial defects, ulceration, corneal vascularization, chronic inflammation, scarring, and conjunctivalization (conjunctival epithelial ingrowth), with resultant loss of the clear demarcation between corneal and conjunctival epithelium at the limbal region [9, 67]. Chronic instability of the corneal epithelium and chronic ulceration may lead to progressive melting of the cornea with the risk of perforation. Because some of the clinical features of limbal deficiency can also be found in other ocular surface disorders, the pathognomonic feature for limbal deficiency is the existence of conjunctival epithelial phenotypes on the corneal surface, caused by conjunctival epithelial ingrowth onto the corneal surface [9, 19]. The conjunctival phenotype with goblet cells on the corneal surface is pathognomonic and can be confirmed by impression cytology. Other clinical signs of limbal deficiency include corneal haziness, leukocytic stromal infiltration and increased corneal permeability to topical fluorescein. Greater permeability of the ingrown conjunctival epithelium permits influx of inflammatory cells or fluorescein dye into corneal tissue more readily than the normal corneal epithelium. Limbal deficiency may be localized (with partial or sectorial limbal damage) or complete (with 360° limbal damage). In partial limbal stem cell deficiency, some sectors of the limbal and corneal epithelium remain normal, and conjunctival ingrowth with abnormal whorl or vortex epitheliopathy is present only at the regions devoid of healthy limbal barriers.

Accurate diagnosis of limbal deficiency is crucial as these patients are poor candidates for conventional corneal transplantation alone, and need an appropriate procedure of limbal stem cell transplantation [9, 19].

# Progenitor Cell Therapy and Tissue Engineering for Corneal Reconstruction

Ocular surface diseases with LSCD, such as Stevens-Johnson syndrome, chemical, thermal and radiation injuries, extensive microbial infection, and inherited disorders such as aniridia, are sight threatening and often cause blindness (see review [19]). Transplantation of a corneal limbal graft that contains corneal epithelial stem cells can help to restore vision. Although corneal transplantation has achieved clinical success, there is an increasing shortage of corneal donors worldwide. Corneal tissue engineering is becoming an important discipline that holds great promise for corneal transplantation to treat the blinding corneal diseases [68-70]. The first clinical success of the cultivated corneal epithelial transplantation was noted in a report of two patients published by Pellegrini and colleagues in 1997 [71]. After this initial report, more investigators reported transplantation of in vitro cultivated corneal epithelium on the substrate carriers using amniotic membrane [53, 72, 73], fibrin [74, 75], and collagen hydrogel [76]. Limbal stem cells and optical substrates are known to be the most important factors for corneal tissue bioengineering in regenerative medicine [69].

Our recent study explored the utilization of natural corneal stroma as an optimal substrate to construct a native like corneal equivalent [77]. Human corneal epithelium was cultivated from donor limbal explants on corneal stromal discs prepared by FDA approved Horizon Epikeratome system (Figure 2). The morphology, phenotype, regenerative capacity and transplantation potential were evaluated by hematoxylin eosin and immunofluorescent staining, a wound healing model, and the xeno-transplantation of the corneal constructs to nude mice. An optically transparent and stratified epithelium was rapidly generated on donor corneal stromal substrate and displayed native-like morphology and structure (Figure 3). An epithelium with 5-7 multi-layers was observed in all limbal explant cultures on the corneal stromal discs, which resembles the native-like human corneal epithelial morphology and structure in vivo, as examined by cross-sections of the cultures. The prepared corneal stromal discs contain an intact basement membrane that overlies Bowman's layer after the donors' corneal epithelium has been completely removed. We compared the corneal epithelia generated on the both sides, the basement membrane or stroma side of the donor corneal stroma, as well as on the amniotic membrane, and on the hydrogel. It appeared that the artificial corneal epithelium generated on the basement side of stromal disc showed the best morphology and structure similar to the native cornea structure in vivo. The cells were polygonal in the basal layer and became flattened in superficial layers.

In order to evaluate the phenotype of the epithelium cultivated from limbal explants on corneal stromal discs, we compared its expression pattern of corneal epithelial markers with that by the donor corneal and limbal epithelia. The epithelium displayed a phenotype similar to human corneal epithelium in vivo. The differentiation markers, K3, involucrin and connexin 43, were expressed in full or superficial layers. Interestingly, certain basal cells were immunopositive to antibodies against limbal stem/progenitor cell markers ABCG2 and p63, which are usually negative in corneal epithelium in vivo (Figure 4). It suggests that this bioengineered corneal epithelium shared some characteristics of human limbal epithelium in vivo. In a functional assay, this regenerated engineered epithelium was able to regenerate in 4 days following from a 4mm-diameter wound created by a filter paper soaked with 1 N NaOH.

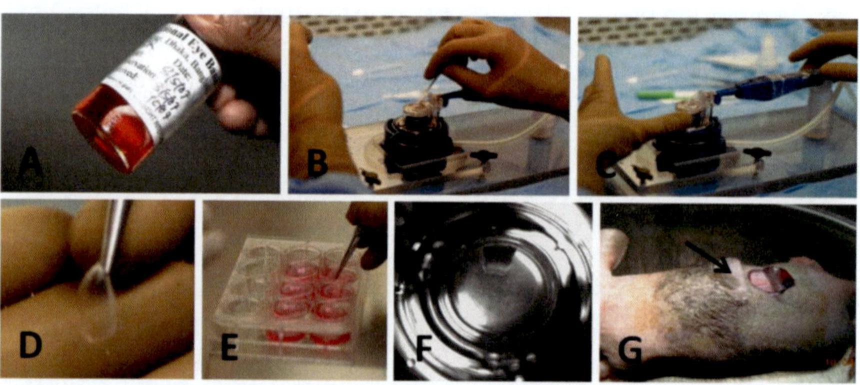

Figure 2. Regeneration of a native-like corneal construct using limbal explants and donor corneal stromal discs. A. Donor corneas stored in Optisol medium; B. Donor corneal epithelium was scraped; C. Horizen Microkeratome system used to fabricate corneal stromal lamella discs; D. A fabricated corneal stromal disc with 10-11mm diameter and 200μm thickness; E. The stroma disc was stored in culture medium; F. A fresh limbal explant cultured on the stromal disc that was placed in the culture insert; G. Xeno-transplantation of the corneal construct into the back skin of a nude mouse.

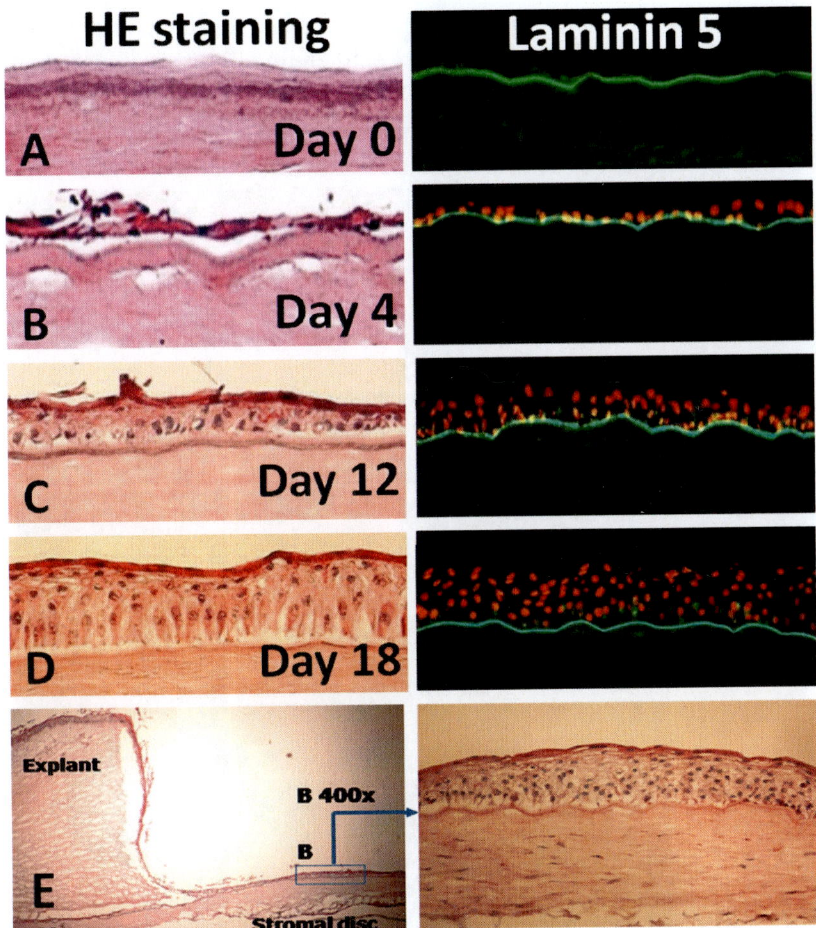



Figure 3. Growth of limbal epithelial cells on corneal stroma. A. The de-epithelial donor corneal stroma contained Bowman's layer with entire basement membrane shown by HE and Laminin 5 staining; B-E. Limbal epithelial cells grew from one to multiple layers from a fresh limbal explant.

In order to evaluate the potential utilization of transplantation, the corneal constructs were xeno-transplanted into the back skin of NIH bg-nudxidBR nude mice. We observed that the epithelium was well survived after xeno-transplantion. The transplanted epithelium remained intact and multilayers after 7 and 14 days as evaluated by HE staining. Immunofluorescent staining showed that corneal epithelial specific marker K3 was expressed by full layers of corneal epithelial cells while epithelial sterm/progenitor cell markers p63, intergrinβ1 and EGFR were strongly immunelocalized at the basal layer cells.

This phenotypic pattern was similar to pre-transplantation condition, suggesting that the transplanted corneal epithelium still shared some features of human limbal epithelium in expression of some progenitor markers. These results suggested that the epithelial cells of corneal construct survived well and still contained progenitor cells after transplantation.

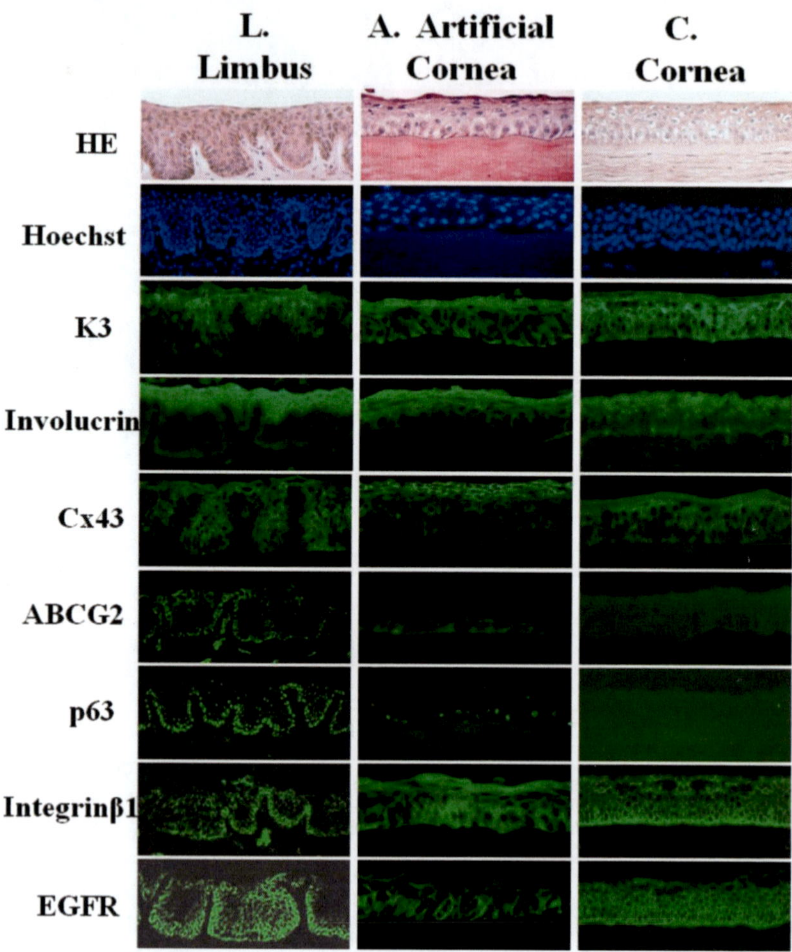



Figure 4. Phenotype of artificial corneal epithelium generated on donor stroma. Representative images of immunofluorescent staining for corneal epithelial markers (Cx43, K3, involucrin, ABCG2, p63, integrin β1 and EGFR) with Hematoxylin-eosin (HE) and Hoechst 33342 nuclear counterstaining on frozen sections of the artificial corneal construct (A), in comparison with those from donor tissues, limbus (L) and cornea (C).

These findings demonstrate that in combination with fresh donor limbal epithelium that contains stem cells, the donor corneal stroma, an optimal source of natural substrate, is useful to bioengineer a native-like corneal equivalent construct with proliferative potential, and a native-like corneal construct has been created with epithelium containing limbal stem/progenitor cells. This corneal construct may have great potential for clinical use, and provide a new approach to corneal reconstruction.

# Conclusion

Stem cells can be defined as cells that have the capacity to self-renew and the ability to generate differentiated progeny or multiple cell lineages. Some adult stem cells may actually represent progenitor cells in that they may turn into cells of the specific tissue in which they reside, but not into any other cell types. Human CESCs represent a great example and model of adult stem or progenitor cells. Human CESCs have been identified to locate in the basal epithelial layer of limbus, and thus also referred as to limbal stem cells. Based on adult stem cell criteria and the characterized phenotype, five clonogenic populations have been isolated, which show the properties that are characteristics of adult stem/progenitor cells. Expansion and cultivation of corneal epithelial progenitor cells have been achieved using different methods, such as limbal tissue explant culture, and limbal epithelial cell suspension co-culture with mouse 3T3 fibroblast feed layer. To avoid the use of xeno-components, two cell lines of commercial human fibroblasts have been identified, which have potential use in replacing mouse 3T3 cells. The concept of CESCs has formed the basis for identifying a class of blinding diseases that display features of corneal epithelial stem cell deficiency or limbal stem cell deficiency. Only transplantation of CESCs can restore vision. Tissue engineering is an important discipline for corneal reconstruction. In combination with fresh limbal epithelium containing stem cells, and the donor corneal stroma, a great source of natural optical substrate, our team has developed a native-like corneal equivalent construct with proliferative potential. This corneal construct provides a new cell therapy for corneal reconstruction.

# References

[1] Watt FM, Hogan BL. Out of Eden: stem cells and their niches. *Science.* 2000;287:1427-1430.

[2] Slack JM. Stem cells in epithelial tissues. *Science.* 2000;287:1431-1433.

[3] Blau HM, Brazelton TR, Weimann JM. The evolving concept of a stem cell: entity or function? *Cell.* 2001;105:829-841.

[4] Lavker RM, Sun TT. Epidermal stem cells: properties, markers, and location. *Proc. Natl. Acad. Sci. USA.* 2000;97:13473-13475.

[5] Cotsarelis G, Kaur P, Dhouailly D, Hengge U, Bickenbach J. Epithelial stem cells in the skin: definition, markers, localization and functions. *Exp. Dermatol.* 1999;8:80-88.

[6] Diaz-Flores L, Jr., Madrid JF, Gutierrez R, Varela H, Valladares F, varez-Arguelles H, Diaz-Flores L. Adult stem and transit-amplifying cell location. *Histol. Histopathol.* 2006;21:995-1027.

[7] Moore KA, Lemischka IR. Stem cells and their niches. *Science.* 2006;311:1880-1885.

[8] Dua HS, Azuara-Blanco A. Limbal stem cells of the corneal epithelium. *Surv. Ophthalmol.* 2000;44:415-425.

[9] Tseng SC. Concept and application of limbal stem cells. *Eye.* 1989;3 ( Pt 2):141-157.

[10] Stepp MA, Zieske JD. The corneal epithelial stem cell niche. *Ocul. Surf.* 2005;3:15-26.

[11] Davanger M, Evensen A. Role of the pericorneal papillary structure in renewal of corneal epithelium. *Nature.* 1971;229:560-561.

[12] Schermer A, Galvin S, Sun T-T. Differentiation-related expression of a major 64K corneal keratin in vivo and in culture suggests limbal location of corneal epithelial stem cells. *J. Cell Biol.* 1986;103:49-62.

[13] Cotsarelis G, Cheng SZ, Dong G, Sun TT, Lavker RM. Existence of slow-cycling limbal epithelial basal cells that can be preferentially stimulated to proliferate: implications on epithelial stem cells. *Cell.* 1989;57:201-209.

[14] Kurpakus MA, Maniaci MT, Esco M. Expression of keratins K12, K4, and K14 during development of ocular surface epithelium. *Curr. Eye Res.* 1994;13:805-814.

[15] Ebato B, Friend J, Thoft RA. Comparison of limbal and peripheral human corneal epithelium in tissue culture. *Invest. Ophthalmol. Vis. Sci.* 1988;29:1533-1537.

[16] Pellegrini G, Golisano O, Paterna P, Lambiase A, Bonini S, Rama P, De Luca M. Location and clonal analysis of stem cells and their differentiated progeny in the human ocular surface. *J. Cell Biol.* 1999;145:769-782.

[17] Chen JJ, Tseng SC. Abnormal corneal epithelial wound healing in partial-thickness removal of limbal epithelium. *Invest. Ophthalmol. Vis. Sci.* 1991;32:2219-2233.

[18] Huang AJ, Tseng SC. Corneal epithelial wound healing in the absence of limbal epithelium. *Invest. Ophthalmol. Vis. Sci.* 1991;32:96-105.

[19] Dua HS, Saini JS, Azuara-Blanco A, Gupta P. Limbal stem cell deficiency: concept, aetiology, clinical presentation, diagnosis and management. *Indian J. Ophthalmol.* 2000;48:83-92.

[20] Budak MT, Alpdogan OS, Zhou M, Lavker RM, Akinci MA, Wolosin JM. Ocular surface epithelia contain ABCG2-dependent side population cells exhibiting features associated with stem cells. *J. Cell Sci.* 2005;118:1715-1724.

[21] Pajoohesh-Ganji A, Stepp MA. In search of markers for the stem cells of the corneal epithelium. *Biol. Cell.* 2005;97:265-276.

[22] Collinson JM, Morris L, Reid AI, Ramaesh T, Keighren MA, Flockhart JH, Hill RE, Tan SS, Ramaesh K, Dhillon B, West JD. Clonal analysis of patterns of growth, stem cell activity, and cell movement during the development and maintenance of the murine corneal epithelium. *Dev. Dyn.* 2002;224:432-440.

[23] Schlotzer-Schrehardt U, Kruse FE. Identification and characterization of limbal stem cells. *Exp. Eye Res.* 2005;81:247-264.

[24] Chen Z, de Paiva CS, Luo L, Kretzer FL, Pflugfelder SC, Li D-Q. Characterization of putative stem cell phenotype in human limbal epithelia. *Stem Cells.* 2004;22:355-366.

[25] Hayashi R, Yamato M, Sugiyama H, Sumide T, Yang J, Okano T, Tano Y, Nishida K. N-cadherin is expressed by putative stem/progenitor cells and melanocytes in the human limbal epithelial stem cell niche. *Stem Cells.* 2007;25:289-296.

[26] Zhou S, Schuetz JD, Bunting KD, Colapietro AM, Sampath J, Morris JJ, Lagutina I, Grosveld GC, Osawa M, Nakauchi H, Sorrentino BP. The ABC transporter Bcrp1/ABCG2 is expressed in a wide variety of stem cells and is a molecular determinant of the side-population phenotype. *Nat. Med.* 2001;7:1028-1034.

[27] de Paiva CS, Chen Z, Corrales RM, Pflugfelder SC, Li D-Q. ABCG2 transporter identifies a population of clonogenic human limbal epithelial cells. *Stem Cells.* 2005;23:63-73.

[28] Yang A, Schweitzer R, Sun D, Kaghad M, Walker N, Bronson RT, Tabin C, Sharpe A, Caput D, Crum C, McKeon F. p63 is essential for regenerative proliferation in limb, craniofacial and epithelial development. *Nature.* 1999;398:714-718.

[29] Parsa R, Yang A, McKeon F, Green H. Association of p63 with proliferative potential in normal and neoplastic human keratinocytes. *J. Invest. Dermatol.* 1999;113:1099-1105.

[30] Pellegrini G, Dellambra E, Golisano O, Martinelli E, Fantozzi I, Bondanza S, Ponzin D, McKeon F, De Luca M. p63 identifies keratinocyte stem cells. *Proc. Natl. Acad. Sci. USA.* 2001;98:3156-3161.

[31] Sariola H. The neurotrophic factors in non-neuronal tissues. *Cell Mol. Life Sci.* 2001;58:1061-1066.

[32] De Sousa PA, da Silva SJ, Anderson RA. Neurotrophin signaling in oocyte survival and developmental competence: a paradigm for cellular toti-potency. *Cloning Stem Cells.* 2004;6:375-385.

[33] Qi H, Chuang EY, Yoon KC, de Paiva CS, Shine HD, Jones DB, Pflugfelder SC, Li DQ. Patterned expression of neurotrophic factors and receptors in human limbal and corneal regions. *Mol. Vis.* 2007;13:1934-1941.

[34] Qi H, Li DQ, Shine HD, Chen Z, Yoon KC, Jones DB, Pflugfelder SC. Nerve growth factor and its receptor TrkA serve as potential markers for human corneal epithelial progenitor cells. *Exp. Eye Res.* 2008;86:34-40.

[35] Qi H, Li DQ, Bian F, Chuang EY, Jones DB, Pflugfelder SC. Expression of glial cell-derived neurotrophic factor and its receptor in the stem-cell-containing human limbal epithelium. *Br. J. Ophthalmol.* 2008;92:1269-1274.

[36] Qi H, Shine HD, Li DQ, de Paiva CS, Farley WJ, Jones DB, Pflugfelder SC. Glial cell-derived neurotrophic factor gene delivery enhances survival of human corneal epithelium in culture and the overexpression of GDNF in bioengineered constructs. *Exp. Eye Res.* 2008;87:580-586.

[37] Stepp MA, Zhu L, Sheppard D, Cranfill RL. Localized distribution of alpha 9 integrin in the cornea and changes in expression during corneal epithelial cell differentiation. *J. Histochem. Cytochem.* 1995;43:353-362.

[38] Jones PH, Watt FM. Separation of human epidermal stem cells from transit amplifying cells on the basis of differences in integrin function and expression. *Cell.* 1993;73:713-724.

[39] Watt FM. Epidermal stem cells: markers, patterning and the control of stem cell fate. *Philos. Trans. R. Soc. Lond. B. Biol. Sci.* 1998;353:831-837.

[40] Pajoohesh-Ganji A, Pal-Ghosh S, Simmens SJ, Stepp MA. Integrins in slow-cycling corneal epithelial cells at the limbus in the mouse. *Stem Cells.* 2006;24:1075-1086.

[41] Zieske JD, Wasson M. Regional variation in distribution of EGF receptor in developing and adult corneal epithelium. *J. Cell Sci.* 1993;106 ( Pt 1):145-152.

[42] Liu Z, Carvajal M, Carraway CA, Carraway K, Pflugfelder SC. Expression of the receptor tyrosine kinases, epidermal growth factor receptor, ErbB2, and ErbB3, in human ocular surface epithelia. *Cornea.* 2001;20:81-85.

[43] Michel M, Torok N, Godbout MJ, Lussier M, Gaudreau P, Royal A, Germain L. Keratin 19 as a biochemical marker of skin stem cells in vivo and in vitro: keratin 19 expressing cells are differentially localized in function of anatomic sites, and their number varies with donor age and culture stage. *J. Cell Sci.* 1996;109 ( Pt 5):1017-1028.

[44] Zieske JD, Bukusoglu G, Yankauckas MA. Characterization of a potential marker of corneal epithelial stem cells. *Invest. Ophthalmol. Vis. Sci.* 1992;33:143-152.

[45] Matic M, Petrov IN, Chen S, Wang C, Dimitrijevich SD, Wolosin JM. Stem cells of the corneal epithelium lack connexins and metabolite transfer capacity. *Differentiation.* 1997;61:251-260.

[46] Scott RAH, Lauweryns B, Snead DMJ, Haynes RJ, Mahida Y, Dua HS. E-cadherin distribution and epithelial basement membrane characteristics of the normal human conjunctiva and cornea. *Eye.* 1997;11:607-612.

[47] Liu C-Y, Zhu G, Converse R, Kao CW-C, Nakamura H, Tseng SCG, Mui M-M, Seyer J, Justice MJ, Stech ME, Hansen GM, Kao WW-Y. Characterization and chromosomal localization of the cornea-specific murine keratin gene *Krt1.12. J. Biol. Chem.* 1994;260:24627-24636.

[48] Tong L, Corrales RM, Chen Z, Villarreal AL, de Paiva CS, Beuerman R, Li D-Q, Pflugfelder SC. Expression and regulation of cornified envelope proteins in human corneal epithelium. *Invest. Ophthalmol. Vis. Sci.* 2006;47:1938-1946.

[49] Banks-Schlegel S, Green H. Involucrin synthesis and tissue assembly by keratinocytes in natural and cultured human epithelia. *J. Cell Biol.* 1981;90:732-737.

[50]  de Paiva CS, Pflugfelder SC, Li D-Q. Cell size correlates with phenotype and proliferative capacity in human corneal epithelial cells. *Stem Cells.* 2006;24:368-375.

[51]  Li D-Q, Chen Z, Song XJ, de Paiva CS, Kim HS, Pflugfelder SC. Partial enrichment of a population of human limbal epithelial cells with putative stem cell properties based on collagen type IV adhesiveness. *Exp. Eye Res.* 2005;80:581-590.

[52]  Chen Z, Evans WH, Pflugfelder SC, Li D-Q. Gap junction protein connexin 43 serves as a negative marker for a stem cell-containing population of human limbal epithelial cells. *Stem Cells.* 2006;24:1265-1273.

[53]  Tsai RJ, Li LM, Chen JK. Reconstruction of damaged corneas by transplantation of autologous limbal epithelial cells. *N. Engl. J. Med.* 2000;343:86-93.

[54]  Koizumi N, Cooper LJ, Fullwood NJ, Nakamura T, Inoki K, Tsuzuki M, Kinoshita S. An evaluation of cultivated corneal limbal epithelial cells, using cell-suspension culture. *Invest. Ophthalmol. Vis. Sci.* 2002;43:2114-2121.

[55]  Kim HS, Jun S, X, de Paiva CS, Chen Z, Pflugfelder SC, Li D-Q. Phenotypic characterization of human corneal epithelial cells expanded ex vivo from limbal explant and single cell cultures. *Exp. Eye Res.* 2004;79:41-49.

[56]  De Luca M, Cancedda R. Culture of human epithelium. *Burns.* 1992;18 Suppl 1:S5-10.

[57]  Liu JY, Burg G. An improved organ culture for regeneration of pure autologous keratinocytes from small split-thickness skin specimens. *Dermatology.* 2005;210:45-48.

[58]  van Rossum MM, Schalkwijk J, van de Kerkhof PC, van Erp PE. Immunofluorescent surface labelling, flow sorting and culturing of putative epidermal stem cells derived from small skin punch biopsies. *J. Immunol. Methods.* 2002;267:109-117.

[59]  Kinoshita S, Koizumi N, Nakamura T. Transplantable cultivated mucosal epithelial sheet for ocular surface reconstruction. *Exp. Eye Res.* 2004;78:483-491.

[60]  Sun T-T, Green H. Cultured epithelial cells of cornea, conjunctiva and skin: absence of marked intrinsic divergence of their differentiated states. *Nature.* 1977;269:489-493.

[61]  Wei Z-G, Wu R-L, Lavker RM, Sun T-T. In vitro growth and differentiation of rabbit bulbar, fornix, and palpebral conjunctival

epithelia. Implication on conjunctival epithelial transdifferentiation and stem cells. *Invest. Ophthalmol. Vis. Sci.* 1993;34:1814-1828.

[62] Lindberg K, Brown ME, Chaves HV, Kenyon KR, Rheinwald JG. In vitro preparation of human ocular surface epithelial cells for transplantation. *Invest. Ophthalmol. Vis. Sci.* 1993;34:2672-2679.

[63] Germain L, Auger FA, Grandbois E, Guignard R, Giasson M, Boisjoly H, Guerin SL. Reconstructed human cornea produced in vitro by tissue engineering. *Pathobiology.* 1999;67:140-147.

[64] Miyashita H, Shimmura S, Kobayashi H, Taguchi T, Asano-Kato N, Uchino Y, Kato M, Shimazaki J, Tanaka J, Tsubota K. Collagen-immobilized poly (vinyl alcohol) as an artificial cornea scaffold that supports a stratified corneal epithelium. *J. Biomed. Mater. Res. B Appl. Biomater.* 2006;76:56-63.

[65] Lavker RM, Tseng SC, Sun TT. Corneal epithelial stem cells at the limbus: looking at some old problems from a new angle. *Exp. Eye Res.* 2004;78:433-446.

[66] Nishida K, Kinoshita S, Ohashi Y, Kuwayama Y, Yamamoto S. Ocular surface abnormalities in aniridia. *Am. J. Ophthalmol.* 1995;120:368-375.

[67] Puangsricharern V, Tseng SCG. Cytologic evidence of corneal diseases with limbal stem cell deficiency. *Ophthalmology.* 1995;102:1476-1485.

[68] Selvam S, Thomas PB, Yiu SC. Tissue engineering: current and future approaches to ocular surface reconstruction. *Ocul. Surf.* 2006;4:120-136.

[69] Germain L, Carrier P, Auger FA, Salesse C, Guerin SL. Can we produce a human corneal equivalent by tissue engineering? *Prog. Retin. Eye Res.* 2000;19:497-527.

[70] Nishida K. Tissue engineering of the cornea. *Cornea.* 2003;22:S28-S34.

[71] Pellegrini G, Traverso CE, Franzi AT, Zingirian M, Cancedda R, De Luca M. Long-term restoration of damaged corneal surfaces with autologous cultivated corneal epithelium. *Lancet.* 1997;349:990-993.

[72] Schwab IR, Reyes M, Isseroff RR. Successful transplantation of bioengineered tissue replacements in patients with ocular surface disease. *Cornea.* 2000;19:421-426.

[73] Koizumi N, Inatomi T, Suzuki T, Sotozono C, Kinoshita S. Cultivated corneal epithelial stem cell transplantation in ocular surface disorders. *Ophthalmology.* 2001;108:1569-1574.

[74] Rama P, Bonini S, Lambiase A, Golisano O, Paterna P, De Luca M, Pellegrini G. Autologous fibrin-cultured limbal stem cells permanently restore the corneal surface of patients with total limbal stem cell deficiency1. *Transplantation.* 2001;72:1478-1485.

[75] Rama P, Matuska S, Paganoni G, Spinelli A, De LM, Pellegrini G. Limbal stem-cell therapy and long-term corneal regeneration. *N. Engl. J. Med.* 2010;363:147-155.

[76] Doillon CJ, Watsky MA, Hakim M, Wang J, Munger R, Laycock N, Osborne R, Griffith M. A collagen-based scaffold for a tissue engineered human cornea: physical and physiological properties. *Int. J. Artif. Organs.* 2003;26:764-773.

[77] Lin J, Yoon KC, Zhang L, Su Z, Lu R, Ma P, de Paiva CS, Pflugfelder SC, Li DQ. A native-like corneal construct using donor corneal stroma for tissue engineering. *PLoS One.* 2012;7:e49571.

In: Progenitor Cells                    ISBN: 978-1-62808-994-3
Editors: P. M. Horton, B. E. Lawrence  © 2013 Nova Science Publishers, Inc.

*Chapter 3*

# Genetically Engineered Blood Pharming: Generation of HLA-Universal Platelets Derived from CD34+ Progenitor Cells

*Constança Figueiredo and Rainer Blasczyk*
Institute for Transfusion Medicine, Hannover Medical School,
Hannover, Germany

## Abstract

Blood pharming is a recently designed concept to enable in vitro production of blood cells that are safe, effective and readily available. This approach represents an alternative to blood donation and may contribute to overcome the shortage of blood products. However, the high variability of the human leukocyte antigen (HLA) loci remains a major hurdle to the application of off-the-shelf blood products. Refractoriness to platelet (PLT) transfusion caused by alloimmunization against HLA class I antigens constitutes a relevant clinical problem. Thus, it would be desirable to generate PLT units devoid of HLA antigens. To reduce the immunogenicity of cell-based therapeutics, we have permanently reduced HLA class I expression using an RNA interference strategy. Furthermore, we demonstrated that the generation of HLA class I-silenced (HLA-universal) PLTs from CD34+ progenitor cells using an shRNA targeting β2-microglobulin transcripts is feasible. CD34+ progenitor cells derived from G-CSF mobilised donors were transduced with a lentiviral vector encoding for the β2-microglobulin-specific shRNA and differentiated into PLTs using a liquid culture system. The functionality of HLA-silenced PLTs and their ability to escape HLA antibody-mediated cytotoxicity were evaluated *in vitro* and *in vivo*. Platelet activation in response to ADP and thrombin were assessed *in vitro*. The immune-evasion capability of HLA-universal megakaryocytes (MKs) and PLTs was tested in lymphocytotoxicity assays using anti-HLA antibodies. To assess the functionality of HLA-universal PLTs *in vivo*, HLA-silenced MKs were infused into NOD/SCID/IL-2Rγc$^{-/-}$ mice with or without anti-HLA antibodies. PLT generation was evaluated by flow cytometry using anti-CD42a and CD61 antibodies. HLA-universal PLTs demonstrated to be functionally similar to blood-derived PLTs. Lymphocytotoxicity assays showed that HLA-silencing efficiently protects MKs against HLA antibody-mediated complement-dependent cytotoxicity. 80-90% of HLA-expressing MKs, but only 3% of HLA-silenced MKs were lysed. *In vivo*, both HLA-expressing and HLA-silenced MKs showed human PLT production (up to 0.5% within the PLT population) when anti-HLA antibodies were absent. However, in presence of anti-HLA antibodies HLA-expressing MKs were rapidly cleared from the circulation of mice, while HLA-silenced MKs escaped HLA antibody-mediated cytotoxicity and human PLT production was detectable up to 11 days. Our studies show that HLA-silenced PLTs are functional and efficiently protected against HLA antibody-mediated cytotoxicity. In this chapter, we provide a review of our most recent findings in the use of CD34+ progenitor cells for the production of HLA-universal PLTs and their potential clinical applications. Provision of HLA-universal PLT units may become an important component in the management of patients with PLT transfusion refractoriness.

# Introduction

Blood pharming is described as the safe and effective production of blood cells. This interdisciplinary concept opened new perspectives in the field of transfusion medicine as it may overcome the shortage in the blood products availability. Blood pharming emerged from the convergence of different research areas such as of stem cell biology, differentiation, hematopoiesis, development of cell culture media and bioreactors. Stem or progenitor cells are the basis for the in vitro generation of blood products.(DARPA; Kim and Baek 2012; Liu et al. 2006; Nielsen 1999; Ratcliffe et al. 2012) However, the use of allogeneic blood products is often associated with a risk for a disadvantageous immune response due to incompatible antigens such as at the human leukocyte antigen (HLA) loci or human platelets antigens (HPA). Use of matched blood products or desensitization protocols have been applied to circumvent those harmful reactions.(Claas et al. 1981; Pavenski et al. 2012; Xia et al. 2012) Both protocols are time or cost consuming and although they have proven some efficacy to improve the use of allogeneic erythrocytes their efficiency in minimizing risks in platelet transfusion remains controversial.(Pavenski et al. 2012; Petz et al. 2000; Rebulla 2005) We have designed a strategy to generate universal blood products. Recently, we have reported the feasibility to stably silence HLA class I expression in several cell types using RNA interference (RNAi) technology.(Figueiredo et al. 2007; Figueiredo et al. 2006; Jaimes et al. 2011) Furthermore, we have combined blood pharming technology with cell bioengineering to produce HLA universal platelets derived from CD34+ cells.(Figueiredo et al. 2010) This chapter provides an overview on the production of HLA universal platelets derived from CD34+ hematopoietic progenitor cells.

# Generation of Platelets In Vitro

Platelets are crucial to maintain haemostasis and their adhesive properties play an important role in stopping bleeding. An adequate supply of platelets is necessary to repair the small vascular damage, and to initiate thrombus formation in open vascular injury. Recent studies suggested vital roles for platelets in wound repair, the innate immune response, and metastatic tumor cell biology.(Garraud and Cognasse 2011; Jenne et al. 2013; Weyrich et al. 2003) Thrombocytopenia is a common clinical condition defined by low platelet counts and might have different etiologies.(Drews 2003; Parker 2012) Nowadays, platelet transfusion occurs in a totally donor dependent manner.(Thiagarajan and Afshar-Kharghan 2013) Blood cellular components such as platelets or erythrocytes lack the capacity to proliferate and therefore it is not possible to expand their number using ex vivo expansion protocols for transfusion purposes similar to those developed for the expansion of lymphocytes (e.g. T and NK cells). Therefore, platelet transfusion is completely dependent on blood/platelet donation or on the large scale differentiation of mature platelets from progenitor cells.(Chen et al. 2009; Reems 2011) Stem cells are characterized by the capacity to renew themselves indefinitely. Cellular plasticity is another property of stem cells and it is defined by their capacity to differentiate into any type of cell.(Bonnet 2003)

Platelet donor shortage, instability and short off the shelf life significantly limit their therapeutic and prophylactic use in thrombocytopenic patients. Thus, their use in a donor-independent manner would significantly facilitate the treatment of those patients, and reduce costs for the health care system.

Platelets are generated in the bone marrow from CD34+ cells by a process known thrombopoiesis. Thrombopoietin and its receptor c-Mpl showed to play an essential role in the expansion and differentiation of Megakaryocyte (MK) progenitors. Also, the nuclear factor erythroid-2 (NF-E2) and GATA1 were identified as the major transcription factors involved in maturation and release of blood platelets. The rarity of MK in the bone marrow prevents their use for ex vivo platelet generation.(Chen et al. 2009; Mancini et al. 2012; Takayama et al. 2010) But, production of platelets in vitro was successfully achieved in vitro and this process has been continuously optimized towards higher yield of platelets. Several cell sources have been used for platelet differentiation. CD34+ progenitor cells isolated from peripheral blood or cord blood are the most common cell sources for the in vitro generation of platelets. However, due to the limited availability of these cells, the search of unlimited cell sources has been initiated. Recent reports have described the production of platelets from embryonic stem cells (ESCs) and induced pluripotent stem cells (iPSCs). However, CD34+ hematopoietic stem cells continue to be the most commonly used cell source for the ex vivo generation of platelets. (Matsunaga et al. 2006; Reems 2011; Takayama et al. 2010; Xia et al. 2012) In addition, we and others have previously demonstrated the feasibility to genetically manipulate CD34+ cells and posteriorly differentiate them in adult cell types.

# Platelet Transfusion Refractoriness

Platelet (PLT) transfusions are commonly used as prophylactic or therapeutic treatment of thrombocytopenic bleeding in patients with hematological or oncological disorders (Blajchman et al. 2008; Sharma et al. 2011). An inadequate response to PLT transfusions, known as PLT refractoriness, is associated with the risk of life-threatening hemorrhage and remains a significant clinical problem (Hod and Schwartz 2008; Phekoo et al. 1997). In hematology/oncology patients, PLT refractoriness has been reported in 7 - 34% of cases (Hod and Schwartz 2008; Rebulla 2005). In approximately one third of these patients PLT refractoriness is caused by immune-mediated mechanisms that most frequently implicate human leukocyte antigen (HLA) class I alloantibodies (Hod and Schwartz 2008; Sacher et al. 2003). In the management of alloimmunized patients refractory to random PLT units, the selection of HLA-matched or crossmatched PLTs has shown to improve post-transfusion PLT count increments in the majority of cases (Hod and Schwartz 2008; Wiita and Nambiar 2012). However, both strategies are time-consuming, require significant organizational and financial resources, and even large blood suppliers have occasionally difficulties to identify HLA-matched or crossmatched PLTs for some patients (Rebulla 2005; Wiita and Nambiar 2012).

Furthermore, the HLA type of HLA-matched PLTs is most likely not identical to the recipients HLA type and may still cause PLT refractoriness in highly immunized patients, which may explain why 25 – 40% of HLA-matched PLT transfusions fail to increase PLT blood counts even in the absence of non-immune causes (Lim et al. 2002). In these cases, it would be desirable to have PLT units devoid of HLA class I antigens on the surface. Generally, alloantibodies are thought to mediate PLT destruction by two mechanisms: 1) antibody-mediated phagocytosis and/or 2) antibody-mediated complement-dependent cytotoxicity (CDC) (Handin and Stossel 1974; Hod and Schwartz 2008; Klein and Blajchman 1982; Pavenski et al. 2012). It is assumed that the main mechanism of PLT clearance is the ingestion of antibody-coated PLTs by phagocytic cells through the Fc-receptor (Lim et al. 2002). Both mechanisms lead to the rapid destruction of allogeneic PLTs in refractory patients and hence impede or even abolish the therapeutic effect of PLT transfusion (Hod and Schwartz 2008).

# Isolation and Culture of CD34+ Progenitor Cells

Peripheral blood CD34$^+$ progenitor cells were isolated from apheresis products HLA-typed HPC donors. Informed consent was obtained from all donors as approved by the local ethics committee of Hannover Medical School. CD34+ cell donors were mobiilized with 10 µg/kg body weight recombinant granulocyte-colony stimulating factor (G-CSF; Lenogastrim, Chugai Pharma, London, UK) for the period of 4 days. CD34$^+$ progenitor cells were isolated from the apheresis product using a human CD34 microbead kit (Miltenyi Biotec, Bergisch Gladbach, Germany) and cultured for 24h in RPMI1640 medium (BioWhittaker/Cambrex, Hess. Oldendorf, Germany) supplemented with 5% human AB serum (C.C. Pro, Neustadt, Germany), thrombopoietin (TPO; 100 ng/mL), Fms-like tyrosine kinase 3-ligand (Flt3-L; 100 ng/mL), stem cell factor (SCF; 100 ng/mL), and interleukin (IL)-6 (50 ng/mL) to keep the cells undifferentiated.

# Generation of HLA Universal Platelets

The HLA complex comprises the most polymorphic loci in the entire human genome. Incompatibility at HLA loci among donors and recipients remains a major problem in transplantation and transfusion medicine. Previously, we have developed a strategy to overcome this hurdle which significantly limits the application of cell-based products derived from an allogeneic source. We have used RNA interference (RNAi) technology to silence the expression of HLA class I molecules. The sequence of a short hairpin RNA targeting β2-microglobulin was designed and cloned into a lentiviral vector (pLVTHm/si, kindly provided by D. Trono) containing an RNAi cassette. Viral particles were produced by co-transfection of Human Embryonic Kidney (HEK)293 cells with the shRNA encoding vector and the helper plasmids psPAX2 (gag/pol) and pMD2G encoding for VSVg envelope protein.

After 24 and 48 h, viral supernatants were collected, filtered, and concentrated by ultracentrifugation (30.000g). CD34$^+$ cells were resuspended in culture medium and seeded into fibronectin-coated 24-well plates. Then, they were infected with the shRNA sequence carrying vector. CD34$^+$ progenitor cells expressing GFP as a reporter marker for successful transduction were isolated by a flow cytometry–based cell sorting system. Transduced CD34+ cells were cultivated in the presence of 100 ng/mL TPO for additional 8 - 10 days to allow the differentiation into MKs and the release of PLTs. As expected, shβ2m-expressing MKs showed a downregulation of β2m transcripts up to 91% (p<0.0001) and HLA class I surface expression up to 64%, as compared to shNS-expressing MKs (p<0.0001) (Figure 1). Differentiation of CD34$^+$ cells expressing shNS or shβ2m yielded a mean of 62% and 70% of mature CD41$^+$CD42a$^+$ MKs, respectively, at day 10 of differentiation. In addition, ploidy analyses demonstrated an increase in DNA content of the differentiating MK, which is a typical feature of these cells. More than 28% of both HLA-silenced and non-silenced MKs showed to be polyploid with a DNA content higher than 4n. In addition, shNS- and shβ2m-transduced MKs produced long pro-PLT extensions starting at day 8 of differentiation similarly to non-transduced MKs (Figure 2, 3). Further cultivation led to a conversion of the MK led cytoplasm into lengthy beaded known as of pro-platelets (pro-PLTs) (Figure 3). This process may be particularly important because mature platelets seem to assemble only at the distal tips of individual proplatelet strands; branching thus represents an elegant mechanism to amplify the number of productive free ends.

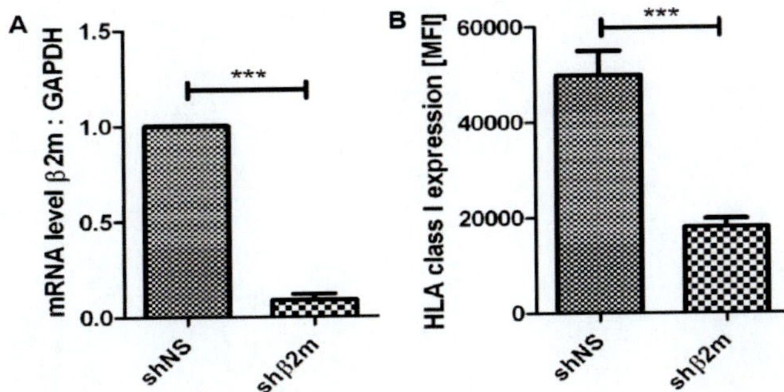


Figure 1. Downregulation of HLA class I expression in CD34[+] progenitor cell-derived MKs. A) Levels of β2m transcripts in shNS- or shβ2m-expressing MKs assessed by real-time PCR. B) HLA class I surface expression of shNS- or shβ2m-expressing MKs analyzed by flow cytometry after staining of cells with the HLA class I-specific antibody w6/32. Both graphs depict means and SD from 3 independent experiments. ***p<0.001.

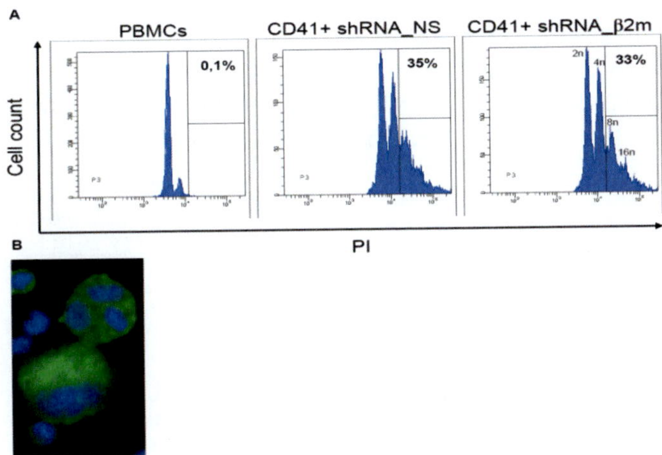


Figure 2. Ploidy analysis of CD34+ progenitor cell-derived HLA class I-silenced MKs. CD34+ progenitor cells expressing shNS or shβ2m were cultured for 10 days in the presence of TPO (100 ng/ml). A) DNA ploidy analysis of shNS- or shβ2m-expressing CD41+ MKs. (B) Representative fluorescence pictures of shβ2m-transduced MKs with polyploid nuclei.

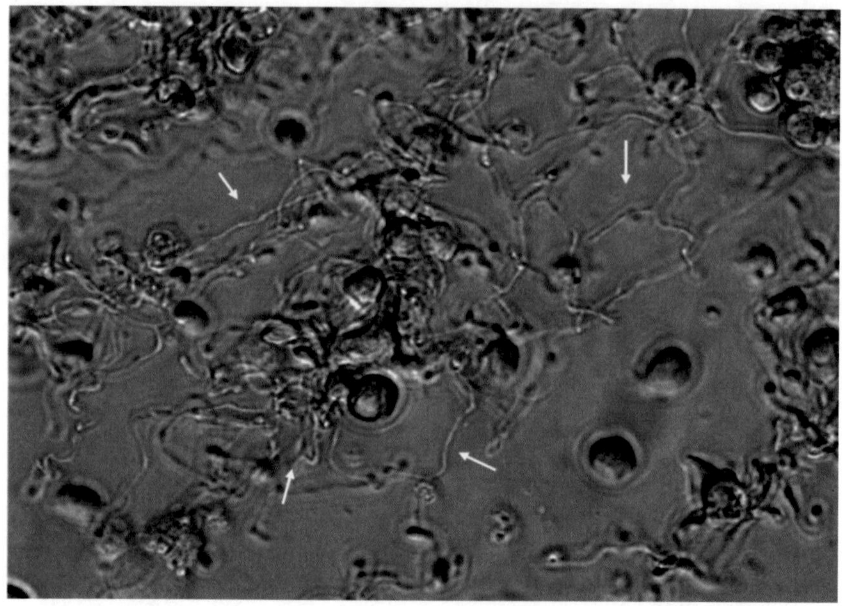

Figure 3. Morphological analyses of pro-PLTs forming MKs. Analysis of non-transduced MKs (NC) and shNS- or shβ2m-expressing MKs (GFP[+]) at day 10 of MK culture using phase contrast and fluorescence microscopy. MKs exhibit long pro-PLTs (indicated by white arrows).

HLA-universal PLTs were characterized by the expression of CD42a and CD61 (GPIIIa) as well as *in vitro* functional assays. At day 12 of differentiation, a mean of 30% of cells in the PLT population derived from shNS- or shβ2m-expressing MKs showed to be CD61[+]CD42a[+].

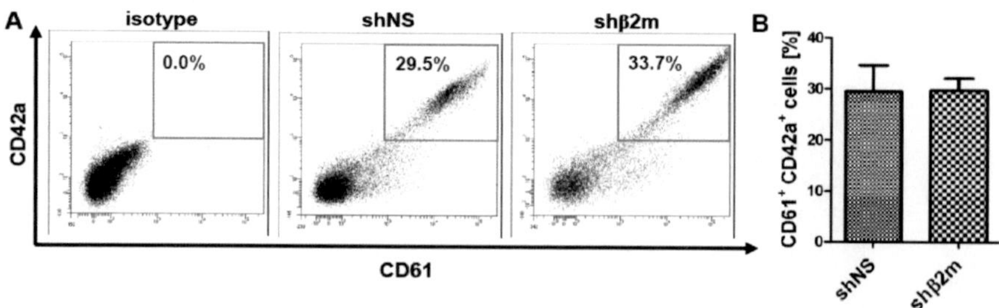


Figure 4. Characterization of CD34[+] progenitor cell-derived HLA class I-silenced PLTs. PLTs derived from shNS- or shβ2m-expressing CD34+ progenitor cells were identified based on FSC/SSC characteristics of blood-derived PLTs and CD42a and CD61 expression at day 12 of differentiation culture. A) Representative flow cytometry dot plots. B) Mean and SD from 4 independent experiments.

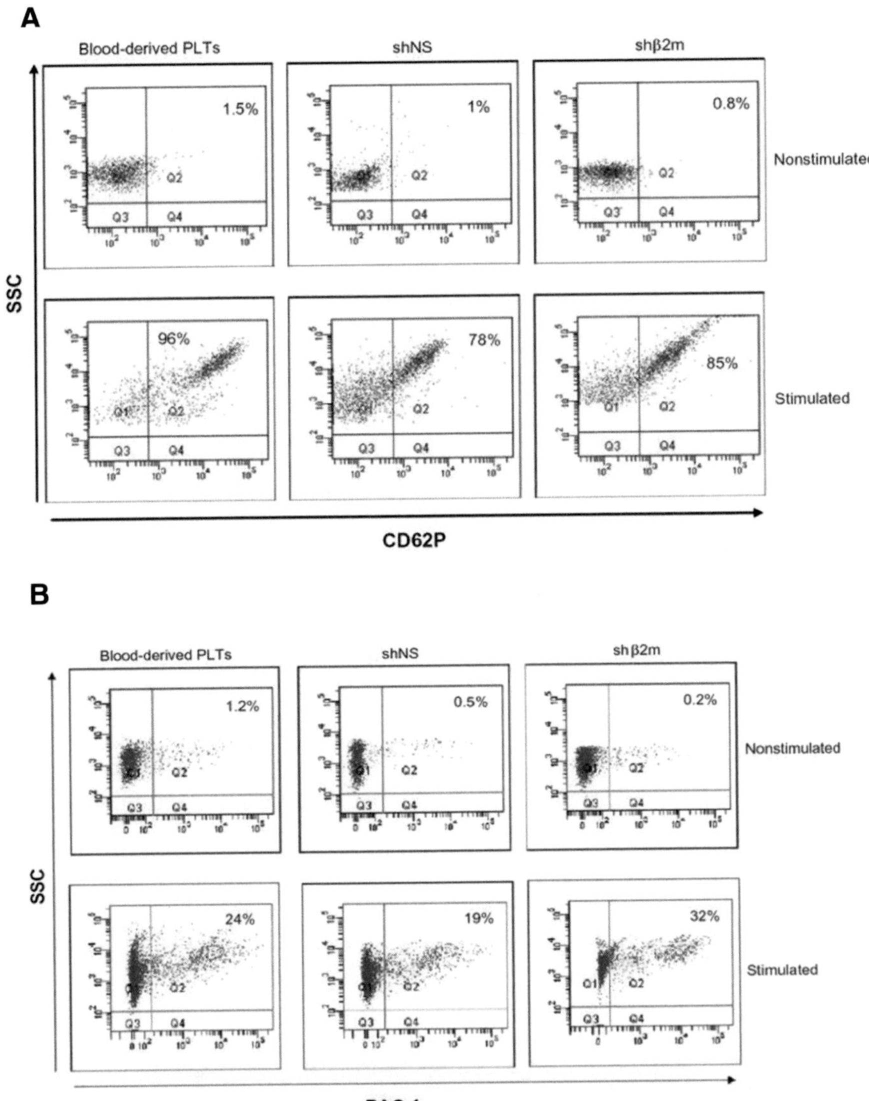


Figure 6. Assessment of activated GPIIb-IIIa and CD62P (P-selectin) expression on CD34+ progenitor cell–derived PLTs after stimulation. Activated GPIIb-IIIa (A) and CD62P (B) were measured in the absence and presence of thrombin (1 U) and ADP (25 mmol/L) on PLTs derived from blood, progenitor cells transduced with the vector encoding for shNS, and progenitor cells transduced with the vector encoding for shb2m. For assessment of activated GPIIb-IIIa the cells were stained with PAC-1, an antibody that mimics the specific fibrinogen binding to human GPIIb-IIIa in the activated state.

For the application of HLA-universal PLTs derived from CD34+ cells as a therapeutic product it is required that they present an equivalent performance to their blood counterparts.

To assess the function of PLTs generated from CD34$^+$ progenitor cells, they were stimulated with 25 mmol/L adenosine 5′-diphosphate (ADP) and 1U thrombin for 10 min and immediately fixed. PLT activation was detected by flow cytometry. PLTs were stained with CD62P and PAC-1 antibodies specific for activated GPIIb-IIIa. HLA-universal PLTs showed to be functional *in vitro,* as shown by upregulation of CD62P and PAC-1. The expression of these markers on the CD34+ progenitor cell–derived PLTs was clearly evoked by this treatment in a fashion similar to those from healthy volunteers (Figure 6).

In particular, there was no difference in ADP- and thrombin-induced activation between the HLA ClassI–silenced and the nonsilenced PLTs. A mean of 75% of HLA Class I–silenced PLTs up regulated CD62P expression and the expression of activated GPIIb-IIIa was increased by up to 35% on these PLTs. Activation with ADP and thrombin produced similar values in PLTs derived from both nontransduced CD34+ progenitor cells and PLTs generated from progenitor cells expressing nonspecific shRNAs (CD62P up regulation of 70 ± 27% and PAC-1 binding of 30 ± 17%). Blood-derived PLTs exhibited CD62P expression levels of 82 17% and PAC-1 binding of 26 ± 7% after stimulation.

PLTs are major players in the formation of the thrombus required to stop bleeding. Thus we have evaluated the capacity of HLA-silenced PLTs to be incorporated in a thrombus. Platelets were labeled with fluorescent nanoparticles and incubated with blood. We have observed that HLA class I silenced platelets were successfully incorporated in the forming thrombus (Figure 7).

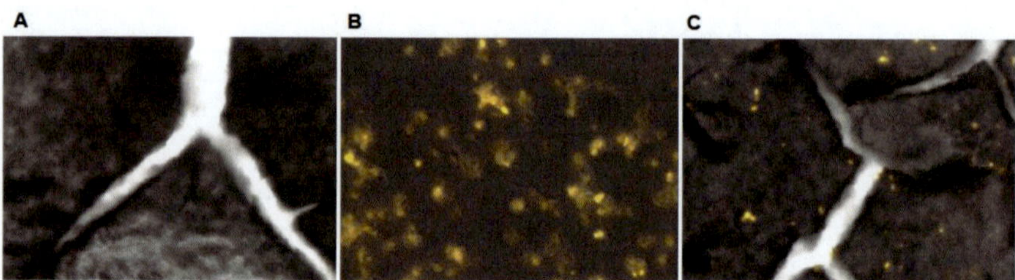

Figure 7. HLA universal platelets are incorporated in the thrombus. (A) Slide preparation of a thrombus (B) HLA class I silenced PLTs labeled with PE-conjugated microparticles (C) HLA class I PLTs incorporated in the thrombus.

# HLA–Universal Megakaryocytes Are Protected from Antibody-Mediated Complement-Dependent Cytotoxicity *In Vitro*

The susceptibility of HLA-universal MKs to undergo anti-HLA antibody-mediated complement-dependent cytotoxicity (CDC) was analyzed using a lymphocytotoxicity assay which is routinely used in clinical transfusion settings (Buakaew and Promwong 2011; Levin et al. 2003).

Lymphocytotoxicity tests were used to assess complement dependent cytotoxicity (CDC) *in vitro*. Thus, viable shNS- or shβ2m-expressing MKs were plated in Terasaki trays (Nunc, Roskilde, Denmark) together with complement-binding donor-specific anti-HLA antibodies. Trays were incubated for 30 min at RT, followed by the addition of 5 µl of rabbit complement (Bio-Rad).

After 60 min of incubation at RT, 5 µl of FouroQuench dye (OneLambda) were added to stain and fix cells. After 15 min cell lysis was analyzed using an Olympus IX81 fluorescence microscope (Olympus) with FITC- and TexasRed-filters to detect viable cells (green) and lysed cells (orange). For evaluation, samples were blinded and frequencies of lysed cells were estimated by two independent individuals.

In this assay, cells are incubated with antibodies that mediate CDC upon binding to their specific antigen. As expected, incubation of shNS-expressing MKs with specific anti-HLA antibodies and complement resulted in high cell lysis rates with 87 ± 6% of lysed cells similar to lysis rates observed in positive control samples (97 ± 3% of lysed cells).

Background lysis levels of shNS-expressing MKs were 7 ± 3% and 8 ± 3% in samples with PBS alone or isotype controls, respectively. Incubation of HLA-silenced MKs with PBS (negative control; NC), isotype antibodies or positive control (PC) reagent, led to lysis rates similar to the corresponding non-silenced MK samples (NC: 7 ± 3%; isotype antibodies: 8 ± 3%; PC: 92 ± 3%).

However, when HLA-silenced MKs were incubated in the presence of specific anti-HLA antibodies and complement, cell lysis was completely abrogated in comparison to non-silenced MKs (p < 0.0001) and lysis rates were equivalent to background lysis levels. These data show that HLA-silenced MKs are efficiently protected from antibody-mediated CDC.

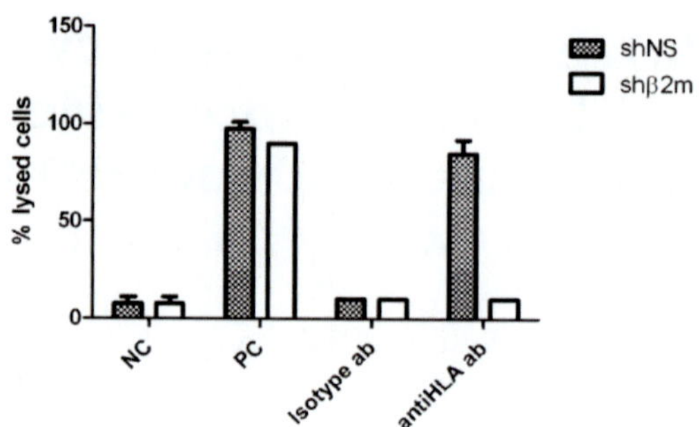


Figure 8. HLA-silenced MKs are protected from anti-HLA antibody mediated complement-dependent cytotoxicity. Lymphocytotoxicity assays were performed with shNS- or shβ2m-expressing MKs (day 10 of differentiation) using complement-fixing anti-HLA antibodies according to the HLA type of HPC donors. A) Representative fluorescence pictures showing viable cells (green) and lysed cells (yellow) after incubation with PBS (NC), positive control reagent (PC), isotype control antibody or specific anti-HLA antibody. B) Mean and SD of 3 independent experiments.

# HLA-Universal Megakaryocytes Increase Human PLT Counts in Mice Refractory to Human PLTS

To assess the capacity of HLA-universal PLTs to escape antibody-mediated cytotoxicity in vivo we used a mouse model for PLT refractoriness. Specific anti-HLA antibodies were injected intravenously (i.v.) into 8 to 10-weeks-old NOD/SCID/IL-2Rγc$^{-/-}$ mice and allowed to distribute within the circulation of mice for 20 min. Subsequently, shNS- or shβ2m-expressing MKs were infused into mice via the tail vein. Mice that received anti-HLA antibody only or MKs only were used as controls (Figure 9).

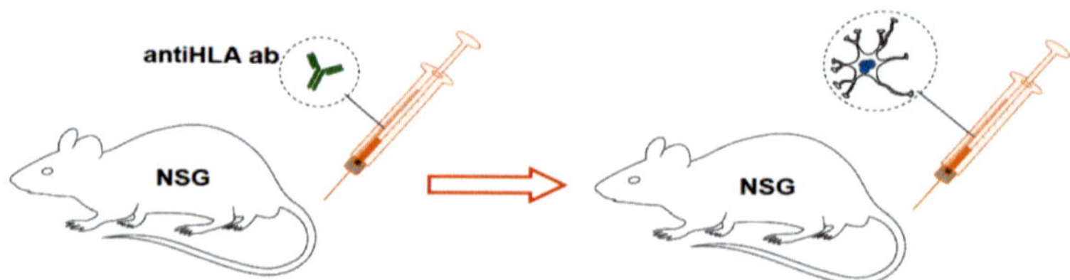


Figure 9. Schematic representation of a platelet refractoriness mouse model. Mice were treated with 3µg/g HLA-specific antibody. After 20 min, 1x106 were non-silenced or HLA silenced MK were infused via the tail vein.

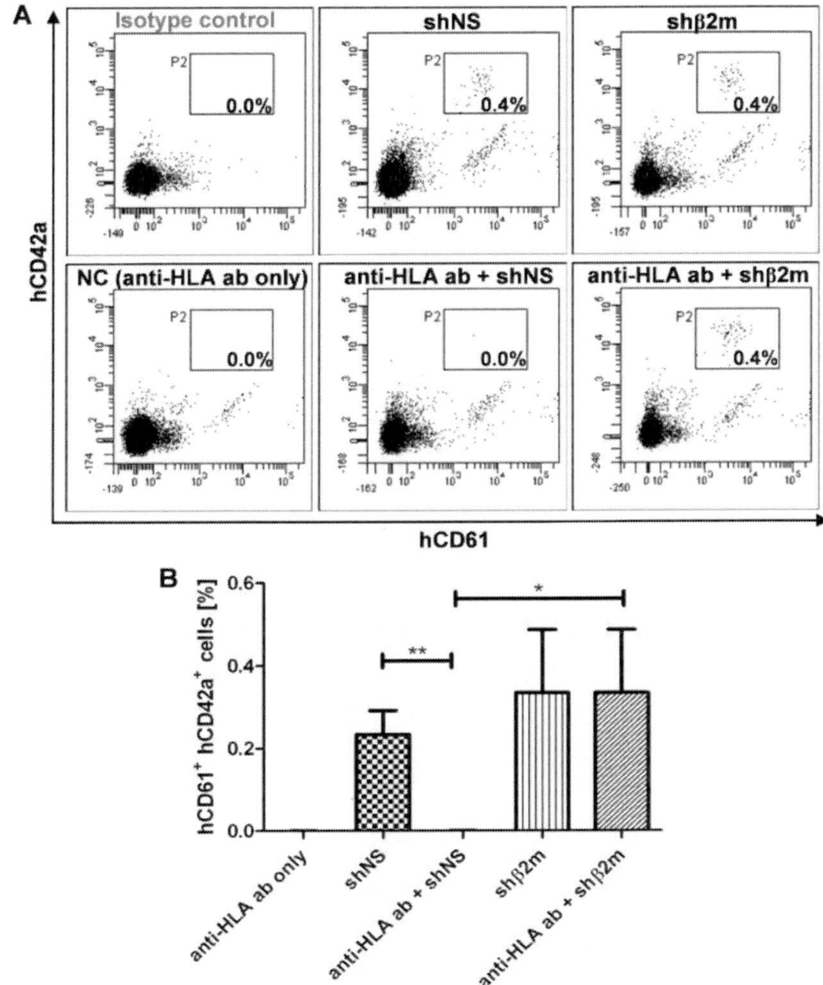


Figure 10. HLA-silenced PLTs escape antibody mediated cytotoxicity in the circulation of living mice. NOD/SCID/IL-2Rγc[-/-] mice were injected i.v. with specific anti-HLA antibodies (to make them refractory to human PLTs) followed by infusion of $1 \times 10^6$ shNS- or shβ2m-expressing MKs. Mice that received antibody only were used as negative control (NC) and mice that received shNS- or shβ2m-expressing MKs only were used as positive control. Mouse PB was harvested 24 h after injection of MKs. Human PLTs were identified based on the co-expression of human (h)CD42a and hCD61. As additional control, PB of mice that received MKs only was stained with isotype controls for anti-hCD42a and anti-hCD61 antibody. A) Representative flow cytometry dot plots. B) Mean + SD from 3 independent experiments. Levels of significance were expressed as p-values (*p<0.05 and **p<0.01).

After 1, 3, 6, and 11 days post transplantation, 40 µl of peripheral blood (PB) was collected from all mice. Human MKs and PLTs in mouse PB were detected using flow cytometry upon staining with anti-human CD41-APC/Cy7 and CD42a-PE or CD42a-PE and CD61-APC, respectively. Twenty-four hours after injection, we observed human PLT ($CD42a^+CD61^+$ cells) frequencies of $0.23 \pm 0.05\%$ and $0.33 \pm 0.15\%$ within the PLT population of mice that received non-silenced or HLA-silenced MKs, respectively. However, no human PLTs were detected in the PLT population of mice that were injected with shNS-expressing MKs together with specific anti-HLA antibodies indicating that human cells were rapidly destroyed in the circulation of these mice. In contrast, injection of HLA-silenced MKs into mice that also received specific anti-HLA antibodies resulted in human PLT frequencies of $0.33 \pm 0.15\%$ ($p < 0.05$) similar to PLT frequencies observed in the absence of anti-HLA antibodies (Figure 9). Human PLTs were detectable in these mice for up to 11 days (data not shown). These results demonstrate that HLA-silenced MKs and PLTs are protected from antibody-mediated cytotoxicity *in vivo* and increase PLT counts in mice refractory to human PLTs.

It is important to note that in all animal experiments, mice behaved normally and did not show any adverse effects up to 2 months after injection of genetically engineered MKs.

# Conclusion

## Universal Cells Are Ideal Blood Products

Platelet transfusions are a standard therapy for patients suffering of thrombocytopenia or platelet dysfunction syndromes. Patients of the surgical or hemato/oncological setting are often dependent on the continuous supply of platelets to prevent hemorrhage which often constitute a threat to patients' life. In particular patients suffering of hematological malignancies repeatedly require platelet transfusions. (Estcourt et al. 2012; Seligsohn 2012; Vassallo 2009; Yu et al. 2013) Due to blood donor scarcity and short storage period of platelets, the large scale in vitro production of platelets from CD34+ progenitor cells would be highly desirable. In contrast to megakaryocytes, CD34+ progenitor cells are easily mobilized into the peripheral blood circulation and collected by apheresis. Recent advances in the development of cell culture techniques and in the knowledge about hematopoiesis in particular in thrombopoiesis allowed to mimic this process in vitro. (Lasky and

Sullenbarger 2011) These progresses have been exploited to develop and improve platelet pharming. Platelet transfusion is usually performed in an allogeneic manner. Also, the application of pharmed platelets will be in an allogeneic manner. Both practices represent a hurdle regarding their immune compatibility. Most of the patients mount an immune response against variable antigens which differ between the donor and recipient pair. Those are mainly human platelet antigens and the high polymorphic HLA. The high level of polymorphism within the HLA locus is the major reason for alloimmunization upon transplantation/transfusion of allogeneic products or following pregnancy (Ayala García et al. 2012). Accordingly, in case of PLT transfusion HLA class I molecules are the major target of antibody-mediated alloimmune responses leading to PLT transfusion refractoriness (Hod and Schwartz 2008; Sacher et al. 2003). Therefore, HLA molecules are an important target to reduce the immunogenicity of *in vitro* generated PLTs. Previously, we have developed a strategy to generate HLA universal cells by genetically engineering the target cells. Recently we have combined this gene therapy approach with the concept of blood pharming to produce HLA class I silenced platelets. An increasing number of studies have demonstrated the feasibility to produce ex vivo platelets derived from CD34+ progenitor cells, ESCs or iPS cells. However, the resultant blood products will be mainly in an allogeneic setting. Hence, as in case of blood-derived PLTs, the use of *in vitro* generated allogeneic PLTs would face the problem of PLT refractoriness in alloimmunized patients. Production of genetically engineered progenitor cells as source for PLT production offers the possibility to generate designer PLTs with low immunogenicity to treat patients requiring multiple transfusions. Furthermore, with the development of bioreactors, significant success has been made regarding large-scale production of PLTs for future clinical application. Previously, we demonstrated the successful *in vitro* production of HPC-derived HLA class I-silenced PLTs which showed to be functionally similar to blood derived PLTs (Figueiredo et al. 2010). A stable reduction of HLA class I surface expression was obtained by using a combination of lentiviral gene transfer and RNA interference technology to target the conserved $\beta2m$ molecule which is an essential part of dimeric HLA class I molecules (Figueiredo et al. 2010). We assessed the capacity of HLA-silenced MKs and PLTs to escape anti-HLA antibody-mediated cytotoxicity, which is the main cause of immune-mediated PLT refractoriness (Sacher et al. 2003). Therefore, we produced HLA-universal MKs and PLTs as previously described (Figueiredo et al. 2010). Morphologic and phenotypic analysis confirmed the successful production of mature MKs and PLTs from HLA-silenced CD34+

progenitor cells. Using platelet activation assays we demonstrated that HLA-universal MKs generate functional PLTs *in vitro*. Moreover, we showed that HLA class I silencing efficiently prevents antibody-mediated PLT destruction *in vitro* and *in vivo*. Alloantibodies are generally thought to cause PLT destruction by either CDC or Fc-receptor-mediated phagocytosis, whereas the latter mechanism is considered to be more relevant (Lim et al. 2002). Using *in vitro* lymphocytotoxicity assays, we showed that HLA-silenced MKs are resistant to anti-HLA antibody-mediated CDC, whereas the majority of non-silenced MKs was lysed. These data strongly suggest that HLA-silenced PLTs derived from HLA-silenced MKs are equally protected from antibody-mediated CDC.

We have used a mouse model for PLT refractoriness to assess the immune evasion capacity of HLA-silenced MKs and PLTs *in vivo*. To mimic PLT refractoriness, NOD/SCID/IL-2Rγc$^{-/-}$ mice were injected i.v. with anti-HLA-antibodies. Transfusion of MK is an alternative strategy to increase PLT counts in thrombocytopenia situations. Previously, it has been shown that mature human MKs infused into the circulation of living mice generate PLTs *in vivo* (Chen et al. 2009). Since *in vitro* PLT production is still less efficient than *in vivo* PLT generation (Reems 2011), we decided to infuse mature MKs into the circulation of mice to allow *in vivo* PLT production. Our data show that within 24 h after MK infusion both non-silenced and HLA-silenced MKs produced similar frequencies of human PLTs in the circulation of mice. However, when non-silenced MKs were infused into mice which also received specific anti-HLA antibodies, no human PLTs or MKs were detected in mouse PB at 24 h post infusion. These data strongly suggest that non-silenced human MKs and PLTs were rapidly destroyed in the circulation of these mice by an anti-HLA antibody mediated mechanism mimicking PLT refractoriness. In contrast, HLA-silenced MKs produced human PLTs in the circulation of mice despite of the presence of specific anti-HLA antibodies. After 24 h, HLA-silenced PLTs were detectable in these mice at frequencies similar to control mice. These results demonstrate that HLA-silenced MKs and PLTs are resistant to PLT refractoriness *in vivo*. Since NOD/SCID/IL-2Rγc$^{-/-}$ mice are deficient in the C5 component of the complement system (Foreman et al. 2011), the rapid antibody-dependent destruction of MKs and PLTs in these mice is likely mediated by Fc-receptor-mediated phagocytosis, which is thought to be the major mechanism of PLT destruction in PLT refractory patients (Lim et al. 2002).

Our studies have demonstrated that HLA-universal MKs and PLTs are efficiently protected from anti-HLA antibody-mediated cytotoxicity (CDC and

Fc-receptor-mediated phagocytosis) and prevent PLT refractoriness *in vivo*. Hence, our approach offers a promising strategy for future clinical PLT supply for patients with immune PLT refractoriness obviating the need for time-consuming and cost-intensive HLA-typing and matching.

In our *in vivo* experiments, mature MKs were infused into mice to circumvent the relatively low rates of *in vitro* PLT production. This strategy may also be useful for future clinical application. Importantly, mice did not show any adverse effects after injection of genetically engineered human cells. Biodistribution analysis revealed that two weeks post infusion there were no residual human cells present in PB, BM, lungs, or spleen of mice, indicating that our MK differentiation cultures contained no potentially cancerogenic precursor cells which could have engrafted and survived for prolonged periods of time. Irradiation can be used to eliminate any residual precursor or transformed cells that are potentially cancerogenic (Gekas and Graf 2010). In addition, platelet gamma- or UVA-irradiation are common techniques to increase clinical safety of transfusion practice (Blajchman et al. 2004). We observed that mature irradiated MKs (30 Gy) produced functional PLTs *in vitro* similarly to non-irradiated controls. These results indicate that mature MKs could be irradiated prior to infusion for clinical purposes.

In addition to the treatment of alloimmunized patients, genetically engineered PLTs could be used to avoid alloimmunization of patients receiving PLT transfusions. This would be of particular interest for patients requiring multiple transfusions and/or patients in need for transplantation of allogeneic cells or tissues. Previous studies have shown that allogeneic platelets *per se* do not stimulate direct T-cell allocytotoxicity (Gouttefangeas et al. 2000), however, it is thought that indirect T cell activation may occur upon antigen presentation of peptides derived from allo-HLA molecules (Pavenski et al. 2012). Reduction in immunogenic molecules (i.e. allogeneic HLA molecules) is likely to reduce the probability of alloimmunization through indirect T cell activation. In the current study, we provide the proof-of-concept for the successful generation of genetically engineered PLTs with reduced immunogenicity. Targeting the conserved β2m molecule via RNAi technology resulted in the efficient downregulation of HLA class I-surface expression by MKs and PLTs and conferred protection against anti-HLA antibody mediated cytotoxicity. Previously we showed that another feasible approach to reduce the expression of HLA class I molecules is direct targeting of HLA class I heavy chains (Figueiredo et al. 2007). This approach would allow the reduction of alloantigen expression by MKs and PLTs and would likely also reduce the risk of alloimmunization by indirect T-cell activation.

In conclusion, we showed that *in vitro* generated HLA-universal PLTs derived from CD34+ progenitor cells are efficiently protected from anti-HLA antibody mediated cytotoxicity and antibody-dependent cellular cytotoxicity *in vivo*. Bioengineered HLA-universal MKs and PLTs may become an important clinical therapeutic strategy for the management of PLT refractory patients in the near future. Moreover, they may constitute the only efficient option for alloimmunized patients where even HLA-matched or cross-matched PLT units fail to increase PLT blood counts.

# References


Ayala García M, González Yebra B, López Flores A, Guaní Guerra E. 2012. The major histocompatibility complex in transplantation. *J. Transplant.* 2012: 842141(doi:10.1155/2012/842141).

Blajchman MA, Goldman M, Baeza F. 2004. Improving the bacteriological safety of platelet transfusions. *Transfus. Med. Rev.* 18(1):11-24.

Blajchman MA, Slichter SJ, Heddle NM, Murphy MF. 2008. New strategies for the optimal use of platelet transfusions. *Hematology Am. Soc. Hematol. Educ. Program*:198-204.

Bonnet D. 2003. Hematopoietic stem cells. *Birth Defects Res. C. Embryo Today* 69(3):219-29.

Buakaew J, Promwong C. 2011. Platelet antibody screening by flow cytometry is more sensitive than solid phase red cell adherence assay and lymphocytotoxicity technique: a comparative study in Thai patients. *Asian Pac. J. Allergy Immunol.* 28(2-3):177-84.

Chen TW, Hwang SM, Chu IM, Hsu SC, Hsieh TB, Yao CL. 2009. Characterization and transplantation of induced megakaryocytes from hematopoietic stem cells for rapid platelet recovery by a two-step serum-free procedure. *Exp. Hematol.* 37(11):1330-1339 e5.

Claas FH, Smeenk RJ, Schmidt R, van Steenbrugge GJ, Eernisse JG. 1981. Alloimmunization against the MHC antigens after platelet transfusions is due to contaminating leukocytes in the platelet suspension. *Exp. Hematol.* 9(1):84-9.

DARPA. Blood Pharming Strategic Thrust. [http://www.darpa.mil/Our_Work/ DSO/Programs/Blood_Pharming.aspx]: [http://www.darpa.mil/Our_Work/ DSO/Programs/Blood_Pharming.aspx].

Drews RE. 2003. Critical issues in hematology: anemia, thrombocytopenia, coagulopathy, and blood product transfusions in critically ill patients. *Clin. Chest Med.* 24(4):607-22.

Estcourt L, Stanworth S, Doree C, Hopewell S, Murphy MF, Tinmouth A, Heddle N. 2012. Prophylactic platelet transfusion for prevention of bleeding in patients with haematological disorders after chemotherapy and stem cell transplantation. *Cochrane Database Syst. Rev.* 5:CD004269.

Figueiredo C, Goudeva L, Horn PA, Eiz-Vesper B, Blasczyk R, Seltsam A. 2010. Generation of HLA-deficient platelets from hematopoietic progenitor cells. *Transfusion* 50(8):1690-701.

Figueiredo C, Horn PA, Blasczyk R, Seltsam A. 2007. Regulating MHC expression for cellular therapeutics. *Transfusion* 47(1):18-27.

Figueiredo C, Seltsam A, Blasczyk R. 2006. Class-, gene-, and group-specific HLA silencing by lentiviral shRNA delivery. *J. Mol. Med.* (Berl) 84(5):425-37.

Foreman O, Kavirayani AM, Griffey SM, Reader R, Shultz LD. 2011. Opportunistic bacterial infections in breeding colonies of the NSG mouse strain. *Vet. Pathol.* 48(2):495-9.

Garraud O, Cognasse F. 2011. Platelet Toll-like receptor expression: the link between "danger" ligands and inflammation. *Inflamm Allergy Drug Targets* 9(5):322-33.

Gekas C, Graf T. 2010. Induced pluripotent stem cell-derived human platelets: one step closer to the clinic. *J. Exp. Med.* 207(13):2781-4.

Gouttefangeas C, Diehl M, Keilholz W, Hornlein RF, Stevanovic S, Rammensee HG. 2000. Thrombocyte HLA molecules retain nonrenewable endogenous peptides of megakaryocyte lineage and do not stimulate direct allocytotoxicity in vitro. *Blood* 95(10):3168-75.

Handin RI, Stossel TP. 1974. Phagocytosis of antibody-coated platelets by human granulocytes. *N. Engl. J. Med.* 290(18):989-93.

Hod E, Schwartz J. 2008. Platelet transfusion refractoriness. *Br. J. Haematol.* 142(3):348-60.

Jaimes Y, Seltsam A, Eiz-Vesper B, Blasczyk R, Figueiredo C. 2011. Regulation of HLA class II expression prevents allogeneic T-cell responses. *Tissue Antigens* 77(1):36-44.

Jenne CN, Urrutia R, Kubes P. 2013. Platelets: bridging hemostasis, inflammation, and immunity. *Int. J. Lab. Hematol.* 35(3):254-61.

Kim HO, Baek EJ. 2012. Red blood cell engineering in stroma and serum/plasma-free conditions and long term storage. *Tissue Eng. Part A* 18(1-2):117-26.

Klein CA, Blajchman MA. 1982. Alloantibodies and platelet destruction. *Semin Thromb Hemost.* 8(2):105-15.

Lasky LC, Sullenbarger B. 2011. Manipulation of oxygenation and flow-induced shear stress can increase the in vitro yield of platelets from cord blood. *Tissue Eng. Part C Methods* 17(11):1081-8.

Levin MD, de Veld JC, van der Holt B, van't Veer MB. 2003. Screening for alloantibodies in the serum of patients receiving platelet transfusions: a comparison of the ELISA, lymphocytotoxicity, and the indirect immunofluorescence method. *Transfusion* 43(1):72-7.

Lim J, Kim Y, Han K, Kim M, Lee KY, Kim WI, Shim SI, Kim BK, Kang CS. 2002. Flow cytometric monocyte phagocytic assay for predicting platelet transfusion outcome. *Transfusion* 42(3):309-16.

Liu Y, Liu T, Fan X, Ma X, Cui Z. 2006. Ex vivo expansion of hematopoietic stem cells derived from umbilical cord blood in rotating wall vessel. *J. Biotechnol.* 124(3):592-601.

Mancini E, Sanjuan-Pla A, Luciani L, Moore S, Grover A, Zay A, Rasmussen KD, Luc S, Bilbao D, O'Carroll D and others. 2012. FOG-1 and GATA-1 act sequentially to specify definitive megakaryocytic and erythroid progenitors. *Embo J.* 31(2):351-65.

Matsunaga T, Tanaka I, Kobune M, Kawano Y, Tanaka M, Kuribayashi K, Iyama S, Sato T, Sato Y, Takimoto R and others. 2006. Ex vivo large-scale generation of human platelets from cord blood CD34+ cells. *Stem Cells* 24(12):2877-87.

Nielsen LK. 1999. Bioreactors for hematopoietic cell culture. *Annu. Rev. Biomed. Eng.* 1:129-52.

Parker RI. 2012. Etiology and significance of thrombocytopenia in critically ill patients. *Crit. Care Clin.* 28(3):399-411, vi.

Pavenski K, Freedman J, Semple JW. 2012. HLA alloimmunization against platelet transfusions: pathophysiology, significance, prevention and management. *Tissue Antigens* 79(4):237-45.

Petz LD, Garratty G, Calhoun L, Clark BD, Terasaki PI, Gresens C, Gornbein JA, Landaw EM, Smith R, Cecka JM. 2000. Selecting donors of platelets for refractory patients on the basis of HLA antibody specificity. *Transfusion* 40(12):1446-56.

Phekoo KJ, Hambley H, Schey SA, Win N, Carr R, Murphy MF. 1997. Audit of practice in platelet refractoriness. *Vox. Sang.* 73(2):81-6.

Ratcliffe E, Glen KE, Workman VL, Stacey AJ, Thomas RJ. 2012. A novel automated bioreactor for scalable process optimisation of haematopoietic stem cell culture. *J. Biotechnol.* 161(3):387-90.

Rebulla P. 2005. A mini-review on platelet refractoriness. *Haematologica* 90(2):247-53.

Reems JA. 2011. A journey to produce platelets in vitro. *Transfusion* 51 Suppl 4:169S-176S.

Sacher RA, Kickler TS, Schiffer CA, Sherman LA, Bracey AW, Shulman IA. 2003. Management of patients refractory to platelet transfusion. *Arch. Pathol. Lab. Med.* 127(4):409-14.

Seligsohn U. 2012. Treatment of inherited platelet disorders. *Haemophilia* 18 Suppl 4:161-5.

Sharma S, Sharma P, Tyler LN. 2011. Transfusion of blood and blood products: indications and complications. *Am. Fam. Physician* 83(6):719-24.

Takayama M, Fujita R, Suzuki M, Okuyama R, Aiba S, Motohashi H, Yamamoto M. 2010. Genetic analysis of hierarchical regulation for Gata1 and NF-E2 p45 gene expression in megakaryopoiesis. *Mol. Cell Biol.* 30(11):2668-80.

Thiagarajan P, Afshar-Kharghan V. 2013. Platelet transfusion therapy. *Hematol. Oncol. Clin. North Am.* 27(3):629-43.

Vassallo RR. 2009. Recognition and management of antibodies to human platelet antigens in platelet transfusion-refractory patients. *Immunohematology* 25(3):119-24.

Weyrich AS, Lindemann S, Zimmerman GA. 2003. The evolving role of platelets in inflammation. *J. Thromb. Haemost.* 1(9):1897-905.

Wiita AP, Nambiar A. 2012. Longitudinal management with crossmatch-compatible platelets for refractory patients: alloimmunization, response to transfusion, and clinical outcomes (CME). *Transfusion* 52(10):2146-54.

Xia WJ, Ye X, Tian LW, Xu XZ, Chen YK, Luo GP, Bei CH, Deng J, Santoso S, Fu YS. 2012. Establishment of platelet donor registry improves the treatment of platelet transfusion refractoriness in Guangzhou region of China. *Transfus. Med* 20(4):269-74.

Yu QH, Shen YP, Ye BD, Zhou YH. 2013. Successful use of rituximab in platelet transfusion refractoriness in a multi-transfused patient with myelodysplastic syndrome. Platelets.

In: Progenitor Cells
Editors: P. M. Horton, B. E. Lawrence © 2013 Nova Science Publishers, Inc.
ISBN: 978-1-62808-994-3



# Regulation of Neural Progenitor Cells by Wnt5a-Signaling in the Developmental Central Nervous System

*Mitsuharu Endo[*] and Yasuhiro Minami[*]*
Department of Physiology and Cell Biology, Kobe University,
Graduate School of Medicine, Japan

## Abstract

In the developing central nervous system (CNS), many different types of neurons are generated in an age- and brain region-dependent manner. Thus, developmental neurogenesis must be precisely controlled in order to ensure the well-ordered generation of the appropriate numbers of specialized neurons. Neurons are generally produced from neural progenitor cells (NPCs). Recent findings have indicated that neuronal subtypes are determined partly in NPCs prior to their differentiation. In addition, the number of neurons generated from NPCs mainly depends on a balance between proliferation and differentiation of NPCs, which is regulated by both intrinsic and extrinsic factors. The Wnt-family of secreted glycoproteins, well documented extrinsic factors in the CNS, have been shown to play crucial roles in regulating developmental

---

[*] Correspondence: mendo@med.kobe-u.ac.jp and minami@kobe-u.ac.jp.

neurogenesis. Wnt signaling can be largely classified into β-catenin-dependent (canonical) and -independent (non-canonical) pathways, which are mediated by binding of different Wnt ligands and their cognate receptors, including Frizzled (Fzd), LRP5/6, Ryk and Ror1/2.

In this chapter, we focus on the function of Wnt5a-induced signaling in the regulation of NPCs during developmental neurogenesis. Wnt5a can elicit non-canonical Wnt signaling to regulate diverse cellular functions. We have recently shown that Wnt5a acts to enhance neurogenesis in the cerebral cortex through the maintenance of proliferative and neurogenic NPCs through its binding to the Ror-family of receptor tyrosine kinases, Ror1 and Ror2. In the developing cerebral cortex, glutamatergic neurons are generated from NPCs in the ventricular zone through the production of intermediate progenitor cells (IPCs), which can divide prior to differentiation into neurons to expand the population of neurons. Inhibition of Wnt5a-Ror signaling in NPCs results in a decreased number of IPCs, eventually leading to the reduction of total number of neurons generated from NPCs. In the developing midbrain and olfactory bulb, Wnt5a has been shown to promote the differentiation of NPCs into dopaminergic neurons or GABAergic interneurons, respectively. In both cases, Wnt5a appears to act to specify the differentiation of NPCs into a distinct subtype of neurons, because Wnt5a stimulation increases the number of dopaminergic or GABAergic neurons rather than the total number of neurons. Taken together, these findings indicate that Wnt5a promotes the production of specific neuron subtypes possibly through the regulation of NPCs in different manners, depending on the brain regions. Importantly, the ability of Wnt5a-signaling to provide specific neuronal populations from NPCs as renewable sources would be beneficial for producing replacement cells for therapeutic applications to neurological disorders, i.e. glutamatergic neurons for stroke, dopaminergic neurons for Parkinson's disease, and GABAergic neurons for Huntington's disease.

# 1. Introduction

The brain comprises many distinct regions, each of which is responsible for different functions (e.g., memory, judgment and movement). Each region of the brain contains complex circuits made up of many kinds of neurons. Importantly, almost all neurons are generated during developmental stages through the differentiation from neural progenitor cells (NPCs) in a brain region-specific manner. In the adult mammalian central nervous system (CNS), including the brain and spinal cord, regeneration after injury is limited, at least partly, due to the restricted activity of neurogenesis under the mature CNS environments, although neurons have been shown to be continuously produced in several small parts of the adult mammalian brain [1, 2]. However, recent studies have shown that some insults, including ischemia, can induce neurogenesis even in the physiologically non-neurogenic regions [3-6], indicating that some regenerative capacity is preserved even in the adult mammalian brain. Based on these findings, two major approaches have recently been taken into account for cell replacement therapies to patients with neurodegenerative disorders. One is activation of the regenerative capacity of endogenous NPCs in the adult CNS, and another is cell transplantation therapy [7]. To develop these therapeutic strategies, and apply them for the treatment of damaged specific neurons in each brain region or spinal cord tissue, it is of great importance to gain a deeper understanding of the signals and mechanisms that regulate the generation of specific neuronal populations from NPCs.

Differentiation of NPCs into neurons is regulated primarily by basic helix-loop-helix (bHLH) proneural transcription factors, including *Mash1*, *Math*, *Neurogenin (Ngn)*, and *NeuroD*. These factors promote cell cycle exit and subsequent neuronal differentiation in diverse progenitor populations [8, 9]. The *Hes* bHLH repressor genes, which are known to be downstream targets of Notch signaling, plays an essential role in maintaining NPCs by antagonizing bHLH proneural genes [9, 10]. Therefore, the balance between expression of the repressor-type and the activator-type (proneural) bHLH genes determines the fate of NPCs, i.e. maintenance or differentiation of NPCs. It should be noted that the proneural genes *Mash1* and *Ngn2* induce expression of Notch ligands such as *Delta-like 1 (Dll1)*, which in turn activate Notch signaling in neighboring cells [11]. This inter-cellular regulation, called lateral inhibition, results in production of a diversity of cell types from apparently equivalent cells. The intrinsic mechanism regulating fate determination of NPCs can be controlled spatio-temporally by extrinsic factors, such as FGF (fibroblast growth factor), BMP (bone morphogenic protein), Shh (sonic hedgehog), and Wnt, that coordinates the timing of neurogenesis and number of generated neurons as well as specification of neuronal subtypes in each brain region. Wnt signaling has been implicated in a variety of cellular responses, including proliferation, differentiation, survival, and cell motility during the development of organs and tissues [12, 13]. Interestingly, Wnt signaling also plays crucial roles in regulating neurogenesis in the developing CNS [14, 15].

Signal transduction, elicited by Wnt-family of secreted proteins, can be classified into β-catenin-dependent canonical and -independent non-canonical Wnt signalings [16]. There are 19 Wnt-family members in mammals, and different Wnts can activate different signaling pathways [17]. For example, while Wnt3a primarily activates β-catenin-dependent canonical Wnt signaling, Wnt5a can activate β-catenin-independent non-canonical Wnt signaling, which contains the planar cell polarity (PCP) pathway and the $Ca^{++}$ pathway. Frizzleds (Fzds), LRP5/6, Ror1/2, and Ryk have been shown to act as receptors or co-receptors for Wnt proteins, and the difference in these Wnt-binding receptors expressed on cell surface has been shown to be attributable partly to the types of Wnt signaling pathway(s) activated in a particular cell [17]. The most well documented Wnt signaling in neurogenesis is canonical Wnt signaling, which plays essential roles in regulating proliferation and neuronal specification of NPCs [15, 18-20]. β-catenin-dependent canonical Wnt signaling is elicited by binding of Wnt with Fzds and LRP5/6 receptor complexes, and induces specific gene expression related to cell proliferation or differentiation via the stabilization and nuclear translocation of β-catenin. In contrast to LRP5/6, the Ror-family of receptor tyrosine kinases, Ror1 and Ror2, has been shown to activate non-canonical Wnt signaling by acting as receptors or co-receptors with Fzd for Wnt5a [21, 22]. Recent findings revealed that Wnt5a also regulates neurogenesis in a β-catenin-independent manner. Although Wnt signaling, as well as FGF, BMP, and Shh signalings, are implicated in the neural induction and axial patterning of the neural plate to organize brain regionalization at early embryonic stages, in this chapter we will focus on the roles of Wnt5a-signaling in neurogenesis through the regulation of several types of NPCs generated in the respective brain regions.

# 2. Wnt5a-Ror signaling in Neocortical NPCs

The neocortex is composed of six horizontal layers, which contains a unique subset of neurons, including glutamatergic excitatory projection neurons and GABAergic inhibitory interneurons. Diverse subtypes of the glutamatergic neurons are generated sequentially from the NPCs resident in the ventricular zone (VZ) of the dorsal telencephalon, and the birthdate of neurons is intimately associated with layer-specific neuronal identities, which are regulated by the specific transcription factors [23]. The sequential and continuous generation of diverse neurons is also based on the maintenance of neocortical NPCs within the VZ through an asymmetric cell division, generating one NPC and one intermediate progenitor cell (IPC), a neurogenic transient amplifying cell (Figure 1A). Notch signaling has been shown to play an essential role in maintaining NPCs in an undifferentiating state. In fact, in conditional knockout (KO) mice of *Rbpj* gene, which encodes an intracellular mediator of Notch signaling, NPCs in the VZ prematurely differentiate into neurons, and as a result they are depleted [24, 25].

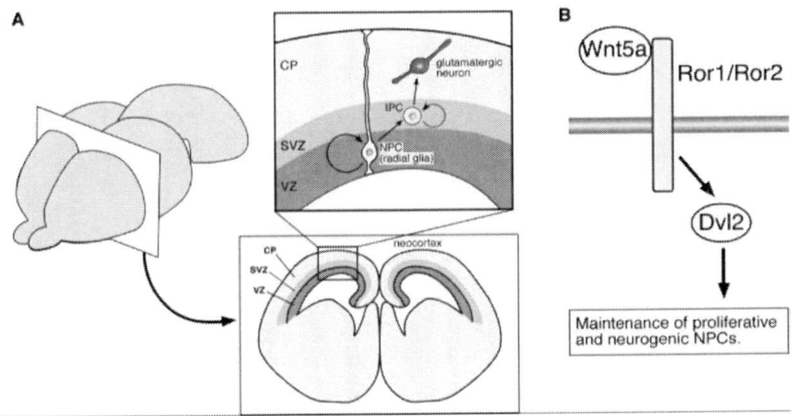


Figure 1. Wnt5a-Ror signaling in neocortical NPCs. A, Radial glia acts as a neural progenitor cell (NPC) that generates an intermediate progenitor cell (IPC) through asymmetric cell division in the ventricular zone (VZ) of the neocortex. IPCs divide prior to differentiation into neurons to expand the population of neurons in the subventricular zone (SVZ). Differentiated neurons migrate into the cortical plate (CP) during their maturation. B, Wnt5a transduces its signal to Dishevelled2 (Dvl2) through Ror1 and Ror2 in neocortical NPCs. Wnt5a-Ror1/Ror2-Dvl2 signaling plays a role in maintaining proliferative and neurogenic NPCs.

We have recently found that Wnt5a-signaling might also be involved in the maintenance of NPCs in an undifferentiating state [26]. Wnt5a and its cognate receptors, Ror1 and Ror2, are significantly expressed in the developing neocortical NPCs. Undifferentiated NPCs, derived from the developing neocortex, can be expanded and maintained *in vitro* by culturing in the presence of basic fibroblast growth factor (bFGF) and/or epidermal growth factor (EGF), and can induce differentiation into IPCs and subsequently neurons by a withdrawal of bFGF and EGF. Small interfering RNA (siRNA)-mediated suppression of *Wnt5a, Ror1*, or *Ror2* in cultured neocortical NPCs results in the reduction of proliferating NPCs as well as IPCs and neurons generated from NPCs under the differentiating condition [26]. Furthermore, suppressed or forced expression of either *Ror1* or *Ror2* in NPCs in the developing neocortex results in the precocious or delayed differentiation of

NPCs into neurons, respectively [26]. Therefore, Wnt5a-Ror1 and Wnt5a-Ror2 signalings might contribute to continuous generation of neurons through the maintenance of proliferative and neurogenic NPCs during neurogenesis of the developing neocortex (Figure 1B). This function of Wnt5a-Ror signaling in NPCs is mediated, at least in part, by Dishevelled 2 (Dvl2), a key mediator of both canonical and non-canonical Wnt signalings, although Wnt5a-Ror signaling appears to act in a β-catenin-independent manner in NPCs [26].

β-catenin-dependent canonical Wnt signaling (Wnt/β-catenin signaling) has also been shown to promote differentiation of neocortical NPCs into IPCs and neurons by inducing expression of Ngn and/or N-Myc [18, 20]. On the other hand, it has been reported that this Wnt/β-catenin signaling promotes self-renewal of NPCs [15, 19]. Although it remains elusive why Wnt/β-catenin signaling, as well as Wnt5a-Ror signaling, can exert opposite effects on NPCs, including the promotion of self-renewal and differentiation of NPCs, Wnt signalings can regulate the self-renewal ability of NPCs in a non-cell autonomous manner through the activation of Notch signaling in surrounding NPCs via the expression of its ligands such as Dll1 in Ngn-expressing IPCs differentiated from NPCs [27-31]. In accordance with this finding, LEF1, a downstream component of Wnt/β-catenin signaling, has been shown to mediate expression of Dll1, thereby activating Notch signaling [32]. Wnts by themselves are not sufficient to support self-renewal of NPCs, because bFGF and/or EGF are also required for self-renewal of NPCs in the cell-culture assays, further indicating that Wnt signalings function cooperatively with other signaling pathways to maintain the self-renewal ability of NPCs.

# 3. Cooperative Function of Wnt5a-signaling with Wnt/β-catenin Signaling in Midbrain NPCs

Parkinson's disease (PD) is a degenerative disorder of the CNS. The motor symptoms of PD result from the death of dopaminergic (DA) neurons in the substantia nigra, a region of the midbrain. It is of importance to establish a reliable method for inducing NPCs into DA neurons to develop possible regenerative therapies for PD, including transplanting NPCs as renewable sources into the degenerated brain and inducing proliferation of endogenous NPCs by pharmacological manipulations [33]. Mesencephalic DA neurons are generated from proliferating NPCs present in the VZ of the ventral midbrain

(Figure 2A). Proliferating NPCs give rise to postmitotic DA precursors that express Nurr1, an orphan nuclear receptor, required for the differentiation of DA precursors into tyrosine hydroxylase-positive (TH+) DA neurons [34]. Similar to the neocortical neurogenesis, proneural genes are crucial regulators that generate DA neurons in the midbrain. Indeed, it has been shown that *Ngn2* is required for the generation of Nurr1+ precursors, and *Mash1* can partially compensate for this function of *Ngn2* [35, 36].

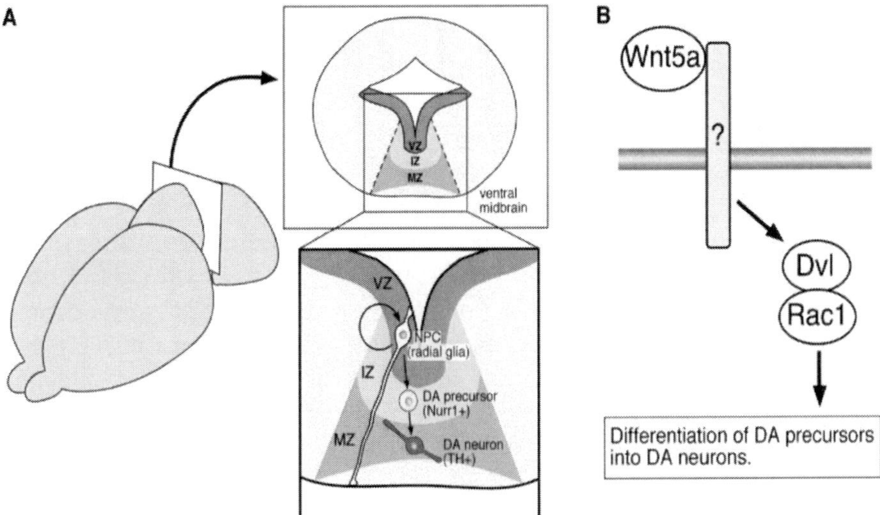


Figure 2. Wnt5a-Rac1 signaling in differentiation of ventral midbrain DA neurons. A, Proliferating neural progenitor cells (NPC), residing in the ventricular zone (VZ) of the ventral midbrain, differentiate into Nurr1-positive (Nurr1+) postmitotic dopaminergic (DA) precursors in the intermediate zone (IZ). Differentiated Nurr1+ precursors further differentiate into tyrosine hydroxylase-positive (TH+) DA neurons in the marginal zone (MZ). B, Wnt5a activates Rac1 via Dishevelled (Dvl) through an unidentified receptor in ventral midbrain NPCs. Wnt5a/Rac1 signaling promotes differentiation of Nurr1+ precursors into TH+ DA neurons.

Several Wnt proteins, including Wnt1, Wnt2, Wnt3, and Wnt5a, are expressed in the developing ventral midbrain [37-39]. Among them, mice, lacking expression of *Wnt1*, exhibit a very severe defect in the midbrain DA neurogenesis [40, 41]. In the mutant mice, expression of both *Ngn2* and *Mash1* is abolished in the floor plate (FP), a most ventral part of the midbrain, and is increased ectopically in the lateral basal plate (BP) [41]. Moreover, TH+ DA neurons are severely reduced in number and are found solely in the lateral BP [40, 41]. In cultured NPCs derived from the ventral midbrain, Wnt1 stimulation increases the proliferation of NPCs, and generation of Nurr1+ precursors and TH+ DA neurons. Wnt1 increases total number of neurons, without affecting the proportion of TH+ DA neurons, indicating that Wnt1 primarily enhances the number of TH+ DA neurons by increasing the proliferation of NPCs [37].

Treatment of ventral midbrain-derived NPCs with Wnt5a *in vitro* also results in robust dose-dependent increases in numbers of TH+ DA neurons [37]. Despite similar effects of Wnt1 and Wnt5a on TH+ DA neurons, Wnt5a appears to increase TH+ DA neurons by promoting differentiation of Nurr1+ precursors into TH+ DA neurons rather than promoting proliferation of NPCs (Figure 2B) [37]. It has been shown that Wnt1 activates Wnt/β-catenin signaling [37], whereas Wnt5a activates Wnt/Rac1 signaling in NPCs of the ventral midbrain [42, 43]. Therefore, it is conceivable that Wnt1 and Wnt5a individually regulate different aspects of the midbrain DA neurogenesis. Although Wnt5a can accelerate TH+ DA neurogenesis *in vitro*, *Wnt5a* KO mice show a transient increase in proliferation of NPCs, a precocious generation of Nurr+ precursors, and a nearly normal number of TH+ DA neurons [42]. In fact, *Wnt5a* KO mice also show a defect in midbrain morphogenesis, in which the ventral midbrain region of DA neurogenesis is expanded laterally and shortened in anterior-posterior axis [42]. This phenotype observed in *Wnt5a* KO mice is reminiscent of a defect in a well-known function of Wnt5a-mediated non-canonical Wnt signaling in convergent extension movements during the developmental morphogenesis. Furthermore, it has recently been shown that Wnt5a cooperates with Wnt1 to promote the generation of midbrain DA neurons [41]. *Wnt1/Wnt5a*-double mutant mice show a more significant loss of Nurr1+ precursors and DA neurons than in *Wnt1*-single mutant mice [41]. This finding further provides a new strategy for the application of Wnts to improve the generation of midbrain DA neurons from NPCs. Embryonic stem (ES) cells have been shown to be capable of producing DA neurons via differentiation into DA NPCs, and sequential administration of Wnt3a, a protein capable of activating Wnt/β-catenin signaling, followed by Wnt5a stimulation significantly increases the proportion of TH+ DA neurons generated from ES cells when compared with control (standard condition) or treatment with Wnt3a alone [41].

# 4. Function of Dlx-induced Expression of *Wnt5a* in GABAergic Interneuron Progenitors of the Olfactory Bulb (OB)

The activity of the large network of principal excitatory neurons in the CNS is balanced by local-circuit interneurons releasing inhibitory neurotransmitters. These interneurons occupy distinct positions within brain microcircuits where they form mainly local connections and perform their functions through a range of various processes. The olfactory bulb (OB) is a structure in the brain that processes information about odors by complex local microcircuitry. This is mediated largely by the actions of a diverse set of interneurons. The OB has a laminar, cortical-like organization. OB interneurons are morphologically and immunohistochemically diverse, and occupy every layer of the OB. At least seven distinct subtypes of OB interneuron populations have been identified according to their position and expression of immune-markers, such as tyrosine hydroxylase (TH), calbindin, calretinin, parvalbumin, and 5T4 [44]. Regardless of these subtypes, most of these OB interneurons express glutamic acid decarboxylase (GAD), GAD65 and/or GAD67, indicating that they are GABAergic inhibitory neurons [45]. OB interneurons are generated beginning from embryogenesis and continuing through adulthood. During the embryonic period, OB interneurons have been shown to arise mainly from NPCs residing in the lateral ganglionic eminence (LGE), whereas postnatally they are derived from the subventricular zone (SVZ) of the lateral ventricle [44, 46]. Development of the OB interneuron is a complex multistep process that involves cell specification in the LGE/SVZ, tangential migration into the OB, and local neuronal maturation there (Figure 3A).

The Dlx-family of homeobox transcription factors, mammalian homologs of *Drosophila* distal-less, play crucial roles in the regulation of migration and differentiation of subpallial NPCs. *Dlx1*, *Dlx2*, *Dlx5*, and *Dlx6* are expressed in the NPCs of OB interneurons. Most, if not all, NPCs of OB interneurons require the expression of *Dlx1* and/or *Dlx2* for their generation and migration [47-49]. *Dlx1/Dlx2*-double mutant mice also lack *Dlx5* expression in most regions of the forebrain [47]. In contrast to the *Dlx1/Dlx2*-double mutant mice, the generation of NPCs in *Dlx5* mutant mice is apparently normal, and its migration from the SVZ of the LGE to the OB also appears to be unaffected [50]. However, *Dlx5* mutant mice have fewer GAD65+, GAD67+, and TH+ neurons in the OB, with apparent accumulation of proliferative and undifferentiated NPCs in the SVZ of the OB [50]. Therefore, Dlx5 appears to be required for NPCs in the SVZ of the OB to mature into postmitotic local circuit interneurons.

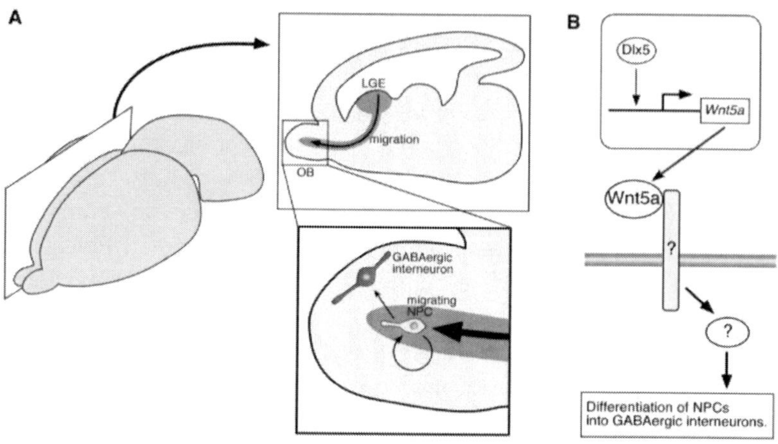


Figure 3. Dlx5-induced expression of Wnt5a in differentiation of GABAergic interneuron. A, Neural progenitor cells (NPCs) originating from the lateral ganglionic eminence (LGE) migrate to the olfactory bulb (OB) to give rise to GABAergic interneurons. B, Dlx5 expressed in the differentiated layers of the OB induces expression of Wnt5a. Wnt5a expressed in the OB promotes differentiation of NPCs into GABAergic interneurons.

It has recently been shown that *Wnt5a* is transcriptionally regulated by the Dlx-family of homeogene transcription factors [51]. In the embryonic OB, *Wnt5a* is expressed in the differentiated layers, where both *Dlx2* and *Dlx5* are expressed, while *Wnt5b*, the closest homolog of *Wnt5a*, is expressed in the VZ-SVZ region, where *Dlx2*, but not *Dlx5*, is expressed [51]. In the OB of *Dlx5* mutant mice, expression of *Wnt5a* is significantly reduced, while expression of *Dlx2* and *Wnt5b* are unaltered [51]. The reduced expression of *Wnt5a* in *Dlx5* mutants might be due to the reduction of its transcription, regulated by the direct binding of Dlx5 to homeodomain binding sites within the *Wnt5a* locus [51]. Moreover, co-culture of OB slices or dissociated OB cells with Wnt5a-expressing cells promotes differentiation of GABAergic interneurons, and this promoting effect of Wnt5a can also be observed even in OB slices or OB cells from *Dlx5* mutant mice where the basal potential of GABAergic differentiation is severely impaired [51]. Importantly, when subpallial NPCs from wild-type (WT) mice are cultured with brain slices or dissociated cells from *Dlx5* mutant mice (*Dlx5* mutant environment), the proportion of GAD67-positive cells is significantly reduced compared to those under the WT environment [51], indicating that Dlx5 promotes differentiation of NPCs into GABAergic interneurons via expression of *Wnt5a* in a non-cell autonomous manner (Figure 3B).

# Conclusion

Many different types of neurons are generated from NPCs within the defined regions of the CNS. In this process, Wnt5a-signaling as well as Wnt/β-catenin signaling play important roles in generating a sufficient number of specific neurons during its own appropriate developmental period. Wnt5a-signaling might regulate proliferation or differentiation of NPCs in a brain region-dependent manner. For example, Wnt5a-signaling maintains proliferative and neurogenic NPCs in the neocortex, whereas it promotes differentiation of NPCs into dopaminergic neurons or GABAergic neurons in the ventral midbrain or OB, respectively. Furthermore, Wnt/β-catenin signaling has been shown to regulate proliferation or differentiation of NPCs, depending on cell types and/or developmental stages. Wnt signaling has been implicated in diverse cellular processes, and one particular Wnt protein can activate different signaling pathways. Variation of Wnt-induced signal transduction is most likely conferred by cellular context as determined by expressed repertoire of receptors and/or signaling molecules rather than by an intrinsic property of the respective Wnt proteins. Although the molecular mechanisms underlying Wnt-induced neurogenesis in each brain region have not fully been understood, different functions of Wnt signaling in each brain region might be due to differences of receptors expressed on NPCs. Furthermore, Wnt5a-signaling appears to regulate NPCs cooperatively with Wnt/β-catenin signaling to generate a sufficient number of specific neuronal populations more efficiently. Therefore, it is of importance to clarify how each Wnt receptor mediates its signaling to regulate neurogenesis in the different types of NPCs.

It has long been thought that adult mammalian CNS does not regenerate after injury. However, recent progress in the field of stem cell research has opened up a way to regenerate damaged CNS by transplanting NPCs as renewable sources. In these replacement therapies, the limited fate determination and poor neurogenic capacity of transplanted NPCs are serious concerns. Thus, it appears necessary to control the differentiation process of NPCs according to the types of neurological disorders, i.e. glutamatergic neurons for stroke, dopaminergic neurons for Parkinson's disease, and GABAergic neurons for Huntington's disease. In this respect, the specific control of Wnt5a-signaling in NPCs would be beneficial to generate efficiently specific neuronal populations for these replacement therapies.

# Acknowledgments

This work was supported by a Grant-in-Aid for Young Scientists (B) (ME) and a Grant-in-Aid for Scientific Research on Innovative Areas (ME, YM) from the Ministry of Education, Culture, Sports, Science and Technology, Japan.

# References

[1] Doetsch F. (2003). The glial identity of neural stem cells. *Nature neuroscience, 6*, 1127-34.

[2] Kriegstein A, Alvarez-Buylla A. (2009). The glial nature of embryonic and adult neural stem cells. *Annual review of neuroscience, 32*, 149-84.

[3] Arvidsson A, Collin T, Kirik D, Kokaia Z, Lindvall O. (2002). Neuronal replacement from endogenous precursors in the adult brain after stroke. *Nature medicine, 8*, 963-70.

[4] Nakatomi H, Kuriu T, Okabe S, Yamamoto S, Hatano O, Kawahara N, et al. (2002). Regeneration of hippocampal pyramidal neurons after ischemic brain injury by recruitment of endogenous neural progenitors. *Cell, 110*, 429-41.

[5] Tonchev AB, Yamashima T, Zhao L, Okano HJ, Okano H. (2003). Proliferation of neural and neuronal progenitors after global brain ischemia in young adult macaque monkeys. *Molecular and cellular neurosciences, 23*, 292-301.

[6] Tonchev AB, Yamashima T, Sawamoto K, Okano H. (2005). Enhanced proliferation of progenitor cells in the subventricular zone and limited neuronal production in the striatum and neocortex of adult macaque monkeys after global cerebral ischemia. *Journal of neuroscience research, 81*, 776-88.

[7] Okano H, Sawamoto K. (2008). Neural stem cells: involvement in adult neurogenesis and CNS repair. *Philosophical transactions of the Royal Society of London Series B, Biological sciences, 363*, 2111-22.

[8] Bertrand N, Castro DS, Guillemot F. (2002). Proneural genes and the specification of neural cell types. *Nature reviews Neuroscience, 3*, 517-30.

[9] Ross SE, Greenberg ME, Stiles CD. (2003). Basic helix-loop-helix factors in cortical development. *Neuron, 39*, 13-25.

[10] Kageyama R, Ohtsuka T, Hatakeyama J, Ohsawa R. (2005). Roles of bHLH genes in neural stem cell differentiation. *Experimental cell research, 306*, 343-8.

[11] Castro DS, Skowronska-Krawczyk D, Armant O, Donaldson IJ, Parras C, Hunt C, et al. (2006). Proneural bHLH and Brn proteins coregulate a neurogenic program through cooperative binding to a conserved DNA motif. *Developmental cell, 11*, 831-44.

[12] Wodarz A, Nusse R. (1998). Mechanisms of Wnt signaling in development. *Annual review of cell and developmental biology, 14*, 59-88.

[13] van Amerongen R, Nusse R. (2009). Towards an integrated view of Wnt signaling in development. *Development, 136*, 3205-14.

[14] Tiberi L, Vanderhaeghen P, van den Ameele J. (2012). Cortical neurogenesis and morphogens: diversity of cues, sources and functions. *Current opinion in cell biology, 24*, 269-76.

[15] Kalani MY, Cheshier SH, Cord BJ, Bababeygy SR, Vogel H, Weissman IL, et al. (2008). Wnt-mediated self-renewal of neural stem/progenitor cells. *Proceedings of the National Academy of Sciences of the United States of America, 105*, 16970-5.

[16] Gordon MD, Nusse R. (2006). Wnt signaling: multiple pathways, multiple receptors, and multiple transcription factors. *The Journal of biological chemistry, 281*, 22429-33.

[17] van Amerongen R, Mikels A, Nusse R. (2008). Alternative wnt signaling is initiated by distinct receptors. *Science signaling, 1*, re9.

[18] Hirabayashi Y, Itoh Y, Tabata H, Nakajima K, Akiyama T, Masuyama N, et al. (2004). The Wnt/beta-catenin pathway directs neuronal differentiation of cortical neural precursor cells. *Development, 131*, 2791-801.

[19] Shimizu T, Kagawa T, Inoue T, Nonaka A, Takada S, Aburatani H, et al. (2008). Stabilized beta-catenin functions through TCF/LEF proteins and the Notch/RBP-Jkappa complex to promote proliferation and suppress

differentiation of neural precursor cells. *Molecular and cellular biology*, *28*, 7427-41.

[20] Kuwahara A, Hirabayashi Y, Knoepfler PS, Taketo MM, Sakai J, Kodama T, et al. (2010). Wnt signaling and its downstream target N-myc regulate basal progenitors in the developing neocortex. *Development*, *137*, 1035-44.

[21] Minami Y, Oishi I, Endo M, Nishita M. (2010). Ror-family receptor tyrosine kinases in noncanonical Wnt signaling: their implications in developmental morphogenesis and human diseases. *Developmental dynamics*, *239*, 1-15.

[22] Green JL, Kuntz SG, Sternberg PW. (2008). Ror receptor tyrosine kinases: orphans no more. *Trends in cell biology*, *18*, 536-44.

[23] Kwan KY, Sestan N, Anton ES. (2012). Transcriptional co-regulation of neuronal migration and laminar identity in the neocortex. *Development*, *139*, 1535-46.

[24] Gao F, Zhang Q, Zheng MH, Liu HL, Hu YY, Zhang P, et al. (2009). Transcription factor RBP-J-mediated signaling represses the differentiation of neural stem cells into intermediate neural progenitors. *Molecular and cellular neurosciences*, *40*, 442-50.

[25] Imayoshi I, Sakamoto M, Yamaguchi M, Mori K, Kageyama R. (2010). Essential roles of Notch signaling in maintenance of neural stem cells in developing and adult brains. *The Journal of neuroscience*, *30*, 3489-98.

[26] Endo M, Doi R, Nishita M, Minami Y. (2012). Ror family receptor tyrosine kinases regulate the maintenance of neural progenitor cells in the developing neocortex. *Journal of cell science*, *125*, 2017-29.

[27] Mizutani K, Yoon K, Dang L, Tokunaga A, Gaiano N. (2007). Differential Notch signalling distinguishes neural stem cells from intermediate progenitors. *Nature*, *449*, 351-5.

[28] Kawaguchi D, Yoshimatsu T, Hozumi K, Gotoh Y. (2008). Selection of differentiating cells by different levels of delta-like 1 among neural precursor cells in the developing mouse telencephalon. *Development*, *135*, 3849-58.

[29] Yoon KJ, Koo BK, Im SK, Jeong HW, Ghim J, Kwon MC, et al. (2008). Mind bomb 1-expressing intermediate progenitors generate notch signaling to maintain radial glial cells. *Neuron*, *58*, 519-31.

[30] Kageyama R, Ohtsuka T, Shimojo H, Imayoshi I. (2008). Dynamic Notch signaling in neural progenitor cells and a revised view of lateral inhibition. *Nature neuroscience*, *11*, 1247-51.

[31] Nelson BR, Hodge RD, Bedogni F, Hevner RF. (2013). Dynamic Interactions between Intermediate Neurogenic Progenitors and Radial Glia in Embryonic Mouse Neocortex: Potential Role in Dll1-Notch Signaling. *The Journal of neuroscience, 33*, 9122-39.

[32] Galceran J, Sustmann C, Hsu SC, Folberth S, Grosschedl R. (2004). LEF1-mediated regulation of Delta-like1 links Wnt and Notch signaling in somitogenesis. *Genes & development, 18*, 2718-23.

[33] Meyer AK, Maisel M, Hermann A, Stirl K, Storch A. (2010). Restorative approaches in Parkinson's Disease: which cell type wins the race? *Journal of the neurological sciences, 289*, 93-103.

[34] Smidt MP, Burbach JP. (2007). How to make a mesodiencephalic dopaminergic neuron. *Nature reviews Neuroscience, 8*, 21-32.

[35] Kele J, Simplicio N, Ferri AL, Mira H, Guillemot F, Arenas E, et al. (2006). Neurogenin 2 is required for the development of ventral midbrain dopaminergic neurons. *Development, 133*, 495-505.

[36] Andersson E, Jensen JB, Parmar M, Guillemot F, Bjorklund A. (2006). Development of the mesencephalic dopaminergic neuron system is compromised in the absence of neurogenin 2. *Development 133*, 507-16.

[37] Castelo-Branco G, Wagner J, Rodriguez FJ, Kele J, Sousa K, Rawal N, et al. (2003). Differential regulation of midbrain dopaminergic neuron development by Wnt-1, Wnt-3a, and Wnt-5a. *Proceedings of the National Academy of Sciences of the United States of America, 100*, 12747-52.

[38] Rawal N, Castelo-Branco G, Sousa KM, Kele J, Kobayashi K, Okano H, et al. (2006). Dynamic temporal and cell type-specific expression of Wnt signaling components in the developing midbrain. *Experimental cell research, 312*, 1626-36.

[39] Sousa KM, Villaescusa JC, Cajanek L, Ondr JK, Castelo-Branco G, Hofstra W, et al. (2010). Wnt2 regulates progenitor proliferation in the developing ventral midbrain. *The Journal of biological chemistry, 285*, 7246-53.

[40] Prakash N, Brodski C, Naserke T, Puelles E, Gogoi R, Hall A, et al. (2006). A Wnt1-regulated genetic network controls the identity and fate of midbrain-dopaminergic progenitors in vivo. *Development, 133*, 89-98.

[41] Andersson ER, Salto C, Villaescusa JC, Cajanek L, Yang S, Bryjova L, et al. (2013). Wnt5a cooperates with canonical Wnts to generate midbrain dopaminergic neurons in vivo and in stem cells. *Proceedings of the National Academy of Sciences of the United States of America, 110*, E602-10.

[42] Andersson ER, Prakash N, Cajanek L, Minina E, Bryja V, Bryjova L, et al. (2008). Wnt5a regulates ventral midbrain morphogenesis and the development of A9-A10 dopaminergic cells in vivo. *PloS one, 3,* e3517.

[43] Cajanek L, Ganji RS, Henriques-Oliveira C, Theofilopoulos S, Konik P, Bryja V, et al. (2013). Tiam1 regulates the Wnt/Dvl/Rac1 signaling pathway and the differentiation of midbrain dopaminergic neurons. *Molecular and cellular biology, 33,* 59-70.

[44] Batista-Brito R, Close J, Machold R, Fishell G. (2008). The distinct temporal origins of olfactory bulb interneuron subtypes. *The Journal of neuroscience, 28,* 3966-75.

[45] Parrish-Aungst S, Shipley MT, Erdelyi F, Szabo G, Puche AC. (2007). Quantitative analysis of neuronal diversity in the mouse olfactory bulb. *The Journal of comparative neurology, 501,* 825-36.

[46] Bovetti S, Peretto P, Fasolo A, De Marchis S. (2007). Spatio-temporal specification of olfactory bulb interneurons. *Journal of molecular histology, 38,* 563-9.

[47] Anderson SA, Qiu M, Bulfone A, Eisenstat DD, Meneses J, Pedersen R, et al. (1997). Mutations of the homeobox genes Dlx-1 and Dlx-2 disrupt the striatal subventricular zone and differentiation of late born striatal neurons. *Neuron, 19,* 27-37.

[48] Bulfone A, Wang F, Hevner R, Anderson S, Cutforth T, Chen S, et al. (1998). An olfactory sensory map develops in the absence of normal projection neurons or GABAergic interneurons. *Neuron, 21,* 1273-82.

[49] Long JE, Garel S, Alvarez-Dolado M, Yoshikawa K, Osumi N, Alvarez-Buylla A, et al. (2007). Dlx-dependent and -independent regulation of olfactory bulb interneuron differentiation. *The Journal of neuroscience, 27,* 3230-43.

[50] Long JE, Garel S, Depew MJ, Tobet S, Rubenstein JL. (2003). DLX5 regulates development of peripheral and central components of the olfactory system. *The Journal of neuroscience, 23,* 568-78.

[51] Paina S, Garzotto D, DeMarchis S, Marino M, Moiana A, Conti L, et al. (2011). Wnt5a is a transcriptional target of Dlx homeogenes and promotes differentiation of interneuron progenitors in vitro and in vivo. *The Journal of neuroscience, 31,* 2675-87.


*Reviewed by* Dr. Toru Takumi, M.D., Ph.D., Principal Investigator, CREST, RIKEN Brain Science Institute (BSI), 2-1, Hirosawa, Wako, Saitama 351-0198, Japan

In: Progenitor Cells                                    ISBN: 978-1-62808-994-3
Editors: P. M. Horton, B. E. Lawrence  © 2013 Nova Science Publishers, Inc.



# Chapter 5

# Endothelial Progenitor Cells in Clinical Settings

*Fumihiro Sanada[1], Yoshiaki Taniyama[1,2,*],*
*Junya Azuma[1,2], Ikeda-Iwabe Yuka[1],*
*Masaaki Iwabayashi[1], Hiromi Rakugi[2]*
*and Ryuichi Morishita[1,2]*

[1]Department of Clinical Gene Therapy, Osaka University Graduate
School of Medicine, Suita, Osaka, Japan
[2]Department of Geriatric Medicine and Nephrology, Osaka University
Graduate School of Medicine, Suita, Osaka, Japan

## Abstract

Senescence of cells is associated with shortened or damaged
telomeres and is characterized by permanent exit from the cell cycle and
altered function. Cellular senescence is caused by repeated cell division,
and also conditions of stress including inflammation and reactive oxygen
species can lead to the development of premature senescence. At the
cellular level, proliferative and oxidative-stress induced cell senescence

---

* Corresponding author: Yoshiaki Taniyama MD, Ph.D., Associate Professor, Department of
Clinical Gene Therapy, Osaka University Graduate School of Medicine, 2-2 Yamada-oka,
Suita, Osaka, 565-0871, Tel: +81-6-6879-3406, FAX: +81-6-6879-3409, E-Mail address;
taniyama@cgt.med.osaka-u.ac.jp.

related to a pro-inflammatory state might strongly contribute to age-associated impaired tissue and organ functions. Vascular cells (endothelial cells, vascular smooth muscle cells) and bone marrow-derived endothelial progenitor cells have been repeatedly shown to have pivotal role in the maintenance and regeneration of cardiovascular tissue. Therefore, the molecular mechanisms of vascular cell senescence have been extensively studied. However, therapeutic approaches to prevent cellular senescence in cardiovascular disease (CVD) are still limited.

Hepatocyte growth factor (HGF), vascular endothelial growth factor (VEGF), and fibroblast growth factor (FGF) are all potent angiogenic growth factors in animal models of ischemia, but their therapeutic effects are not the same in animal experiments and clinical trials. A multicenter, double-blind, placebo-controlled phase III clinical trial in Japan and a US phase II clinical trial of HGF gene therapy for critical limb ischemia (CLI) demonstrated a significant improvement in primary end points and an increase in transcutaneous partial pressure of oxygen even after one year compared with placebo, whereas effectiveness of VEGF and FGF treatment for CLI has not yet been shown. Moreover, our recent publication and another researcher demonstrated that HGF acts as an anti-inflammatory cytokine, while VEGF and FGF act as pro-inflammatory cytokine.

This review overviews the outcomes of clinical trials using angiogenic growth factors, which have shown a dramatic effect in several animal studies. Additionally, interventions with HGF aimed at improving the regenerative capacity of stem/progenitor cells and vascular cells by preventing cellular senescence are discussed.

# **Introduction**

The WHO reports that cardiovascular disease (CVD) accounted for 17.3 million deaths in 2008, representing 30% of all global deaths. Moreover, the number of people who die from CVD will increase to 23.3 million by 2030 [1, 2]. Risk factors for CVD such as tobacco use, unhealthy diet and obesity, physical inactivity, high blood pressure, diabetes and raised lipids induce the impairment of vessel repair in addition to vascular damage, leading to the formation of atherosclerosis and ischemia of organs. Recently, some research groups reported that these undesirable changes are partly attributable to vascular cell senescence, biological aging of the cell [3, 4]. The interaction of endothelial cells with monocytes is increased by endothelial cell senescence, resulting in atherosclerosis [5]. Risk factors for CVD induce bone marrow-derived circulating endothelial progenitor cell (EPC) senescence. EPC function is impaired in patients with coronary artery disease [6], which might contribute to less neovascularization. Additionally, an association between senescent VSMC and their impaired function has been described [7, 8]. Thus, prevention of vascular cell senescence together with therapeutic angiogenesis seems to be an ideal treatment for CVD patients.

Both hepatocyte growth factor (HGF) and vascular endothelial growth factor (VEGF) are potent angiogenic growth factors in animal models of ischemia, but their characteristics are not the same in animal experiments and clinical settings. A multicenter, double-blind, placebo-controlled phase III clinical trial in Japan and a US phase II clinical trial of HGF gene therapy for peripheral artery disease (PAD) demonstrated a significant improvement in primary end points or an increase in transcutaneous partial pressure of oxygen even after 1 year compared with placebo [9], while effectiveness of VEGF gene therapy has not yet been shown for PAD patients.

In this context, we first discuss role of vascular cell senescence in CVD, including EPC, and second, the outcomes of clinical trials of angiogenic growth factors, VEGF, FGF, and HGF, in PAD patients. Finally, we have examined the characteristics of each angiogenic growth factor in inflammation, fibrosis, and cellular senescence in the presence of vascular risk factors, which may account for the success or failure of clinical trials.

# Contribution of EPC to Endothelial Cell Repair and Vascular Cell Senescence

The endothelial cell layer of the vasculature is constantly exposed to mechanical and chemical stress that cause dysfunction of endothelial cells (EC), both in patients with risk factor for atherosclerosis and healthy individuals. Therefore, persistent replacement of damaged EC is required for endothelial homeostasis. About 15 years ago, Asahara et al. discovered that EC repair does not totally depend on the proliferation and migration of resident endothelial cells, but also partly on EPC [10]. Following this breakthrough, Hu et al. showed that EPC can home to the site of vascular injury, where they differentiate and integrate into an endothelial cell layer, although mature EC have limited regenerative capacity [11]. These observations motivated researchers to use EPC for clinical therapy. However, the number and function of EPC are negatively correlated with various risk factors for atherosclerosis [12], potentially reducing their angiogenic capacity and contributing to increased CVD incidence. Accordingly, the number of senescent EPC, with permanent exit from the cell cycle and altered function, increases with risk factors for CVD [6]. In the case of endothelium, firm evidence of its senescence was shown using senescence associated b-galactosidase (SA- b-gal) staining. With this method, senescent EC were found to accumulate in rat carotid artery after balloon injury, a model that provokes EC and VSMC proliferation and neointimal formation [13]. Additionally, another laboratory demonstrated that SA-b-gal-positive senescent EC and VSMC overlie atherosclerotic plaques of human arteries [14]. Moreover, several studies demonstrated that the changes in senescent vascular cells (EC and VSMC) in vitro mimic the condition of human atherosclerosis [*15, 16*]. Although details of the mechanism underlying cellular senescence are limited, intervention against vascular senescence seems to be needed to prevent atherosclerosis as well as subsequent ischemic events.

# Angiogenic Cytokines and Clinical Outcome

Angiogenic cytokines have the ability to increase collateral vessels, and ameliorate perfusion and tissue oxygenation, limiting ischemic lesions in rodents [17, 18]. Therefore, they have been expected to improve the function and symptoms of patients with critical limb ischemia and cardiac ischemia. Initially, interest in utilizing angiogenic cytokines for clinical intervention focused on vascular endothelial growth factor (VEGF) and fibroblast growth factor (FGF). Later, the therapeutic potential of hepatocyte growth factor (HGF) has been identified.

# Vascular Endothelial Growth Factor (VEGF)

VEGF has been widely studied for its angiogenic potential in both cancer and tissue ischemia. The VEGF family is composed of VEGF-A to E. Additionally, VEGF-A and B have isoforms (e.g., $VEGFA_{121}$, $VEGF_{165}$) resulting from alternative splicing of mRNA. $VEGF_{165}$ is the most abundant splice variant of VEGF-A [19] and the best characterized of the isoforms. Experimentally, in a hind-limb ischemia model, over-expression of VEGF by plasmid and adenovirus significantly improved tissue perfusion and oxygenation with neovascularization [20, 21]. VEGF over-expression results in mobilization and recruitment of EPC, with subsequent neovascularization [22, 23]. Clinically, VEGF gene therapy shows contentious findings. Phase I clinical trials demonstrated that naked-plasmid-VEGF$_{165}$ gene therapy is safe, and over-expression of this cytokine in skeletal muscle of PAD patients significantly improved wound healing and perfusion (Table 1) [24]. Similarly, Makinen et al. showed that intra-arterial transfer of AdVEGF$_{165}$ significantly increased vascularity, collaterals, compared to the placebo-control group [25]. Contrary to VEGF$_{165}$ administration, Rajagopalan et al. demonstrated that intramuscular injection of the AdVEGF$_{121}$ isoform caused no improvement in claudication, ankle-brachial index, and quality of life in a phase II clinical trial (26). In addition, to date, no phase III clinical trial using VEGF gene transfer has shown improvement. It should also be noted that VEGF therapy was strongly associated with dose-dependent peripheral edema; 60% of patients developed moderate or severe edema in a clinical trial. However, Ehrbar et al. showed that controlling the level of VEGF with special delivery systems ($\alpha$2PI$_{1-8}$-VEGF$_{121}$) results in non-leaky vessels and improved vessel formation more potently than with native VEGF$_{121}$ [27]. We should consider the possibility that slow-release, low-dose VEGF might be effective and feasible in clinical trials.

# Fibroblast Growth Factor (FGF)

There has also been an attempt to use FGF as a therapeutic gene in the treatment of PAD. The FGF family is composed of 22 members that bind to several FGF receptors. Among them, FGF-1 (aFGF), FGF-2 (bFGF), and FGF-4 have been shown to possess angiogenic potential, the ability to enhance blood-vessel formation, migration, proliferation, and differentiation of EC to facilitate angiogenesis with sprouting of capillaries from vessels. The FGF receptor is expressed on SMC, EC, and EPC [28]. Based on these effects, a naked DNA plasmid vector (non-viral FGF vector; NV1FGF) containing human FGF1 has been developed for therapeutic angiogenesis. In an animal model, the capacity to establish a functional vascular network in the ischemic region was proven [29]. A phase I clinical trial enrolling 51 patients showed a marked improvement in ulcer size, claudication, ankle brachial index, and transcutaneous tissue oxygen in the ischemic limb [30]. A subsequent phase II clinical trial (TALISMAN) also showed promising results, with improvement of ulcer healing and decrease incidence of amputation or death [31]. With great hope for FGF gene therapy, a phase III randomized clinical trial (TAMARIS) was performed in patients who were not eligible for revascularization [32]. However, surprisingly, no benefit of NV1FGF was observed with either the primary efficacy endpoint (major amputation or death), or secondary endpoints (minor amputation, skin lesion status, pain index, quality of life, and ankle and toe brachial index). It is noteworthy that administration of FGF has been linked with potential hypertension and membranous nephropathy; these adverse effects have to be carefully considered in determining the dosage and timing of administration of FGF.

## Table 1. Human clinical trials of angiogenic cytokines in PAD

| Trials | Strategy | Phase | Outcom |
|---|---|---|---|
| Baumgartner et al. (24) | $phVEGF_{165}$ | I | Tolerate |
| Makinen et al. (25) | $phVEGF_{165}$ | I | Tolerate |
| RAVE (26) | $AdenoVEGF_{121}$ | II | No improvement of exercise performance or QOL |
| Greiningen (58) | $phVEGF_{165}$ | II | No reduction in amputaion rate |
| Comerota et al. (30) | phFGF-1 | I | Tolerate |
| TALISMAN (31) | phFGF-1 | II | Reduction in amputation rate |
| TAMARIS (32) | phFGF-1 | III | No reduction in amputation rate or death<br>No improvement of QOL or ABI |
| Morishita et al. (59) | phHGF | I/IIa | Tolerate |
| Makino et al. (36) | phHGF | I/IIa | Improvement of ABI<br>Reduction in rest pain and ulcer size up to 2 years |
| TREAT-HGF (35) | phHGF | III | Improvement in rest pain and ABI<br>Reduction in ulcer size |
| HGF-STAT (9) | phHGF | II | Improvement in TcPO2 |
| HGF 0205 trial (34) | phHGF | II | Improvement in TBI and rest pain |

ABI: ankle-brachial index  TBI: toe brachial index  TcPO2: transcutaneous oxygen tension

# Hepatocyte Growth Factor (HGF)

Initially, HGF was discovered as a potent mitogen for hepatocytes. Later, its angiogenic activity, EC proliferation and migration, via tyrosine phosphorylation of its receptor, c-Met, were discovered. c-Met is expressed on EC, SMC, and also EPC [33]. A clinical trial using HGF for the treatment of hind-limb ischemia is particularly fascinating, since at least three randomized placebo-controlled clinical trials using naked human HGF plasmid DNA confirmed healing of ulcer, decrease in rest pain, and increase in transcutaneus oxygen tension [9, 34, 35]. Additionally, our group proved long-term efficacy of HGF gene therapy up to 2 years, showing an increase in ankle-brachial pressure index, and a reduction of rest pain and size of the ulcer [36]. It is worth mentioning that, different from VEGF gene therapy, HGF gene therapy did not show edema as a side effect.

These clinical results of VEGF, FGF, and HGF for PAD patients cause us to think about the initial results of animal experiments. The following paragraph discusses the differences in inflammation, fibrosis and cell senescence among three angiogenic growth factors, which could possibly affect the results of clinical trial.

# HGF, But Not VEGF and FGF, Prevent Inflammation, Cell Senescence, and Tissue Fibrosis

Although HGF was initially recognized as a mitogen for hepatocytes and a scatter factor, a factor that increases mobility, recent studies revealed that the HGF/c-Met system has several roles in different aspects of cellular biology including cytokine production, cell survival, cell differentiation, and proliferation. Thus, HGF is now recognized as a key growth factor in the prevention and attenuation of acute and chronic disease progression in the heart [37], kidney [38], liver [39], and vascular repair [40, 41]. Of HGF's beneficial functions, its anti-inflammatory and anti-fibrotic action differentiates HGF from VEGF and bFGF. Inflammation, characterized by influx of inflammatory response cells, is thought to play a critical role in the initiation and progression of a wide range of chronic diseases, including kidney, heart, and lung diseases [42, 43, 44]. In addition, recent studies have revealed the involvement of inflammation in cellular senescence [45, 46]. IL-6 and IL-8 play an essential role in the initiation and maintenance of cellular senescence. Reactive oxygen species (ROS) cause cellular senescence as well [47]. Inflammatory reactions induce the production of ROS. The reverse sequence of these events is also true. Thus, a connection between inflammation and cellular senescence is evident. It is also noteworthy that most of the animal experiments, including hind-limb ischemia model and acute myocardial infarction model, have been done in the absence of vascular risk factor, which is present in clinical setting.

To elucidate the discrepancy between the HGF and VEGF clinical trial outcomes, we compared the effects of HGF and VEGF on EPC under angiotensin II (Ang II) stimulation, which is a well-known risk factor for atherosclerosis. We also demonstrated that HGF, but not VEGF, attenuated Ang II-induced senescence of EPC through a reduction of oxidative stress by inhibition of the PIP3/rac1 pathway [41]. Potent induction of neovascularization of EPC by HGF, but not VEGF, under Ang II was also confirmed by in vivo experiments using several models, including HGF transgenic mice. Interestingly, HGF and its receptor, c-Met, regulate Ang II signaling by the Ang II-dependent epithelial growth factor receptor (EGFR) down-regulation via the ubiquitin proteasome system [40]. Moreover, HGF down-regulates EGFR expression following administration of lipopolysaccharide (LPS), endothelin-1 (ET-1), and TGF-b, which transactivate EGFR, suggesting that ligand-dependent EGFR down-regulation of HGF might be the major anti-inflammatory mechanism of HGF [40, 48]. Consistent with our finding, Kaga et al. proved that bFGF alone, but not HGF, significantly increased the growth of VSMC, and activated an essential transcription factor for inflammation, NFκB, and gene expression of its downstream inflammation-related cytokines (IL-8 and MCP-1) in VSMC, accompanied by an increase in vascular permeability in a rat paper disc model [49]. Ohtani et al. showed that expression of VEGF increases after neointimal injury, recruiting monocyte-lineage cells [50]. Importantly, HGF was shown to have a synergistic action with VEGF on EC proliferation and chemotactic response and neovascularization [51, 52]. In these studies, the authors also demonstrated that HGF up-regulated VEGF mRNA transcript, although HGF gene therapy did not cause leaky vessels or edema. Min et al. reported that HGF markedly reduced the increase in leukocyte adhesion and adhesion molecule expression stimulated by VEGF. This effect was mediated by HGF suppression of VEGF-induced NFkB signaling [53]. Thus, HGF stimulates new vessel growth without the complication of inflammation, edema, and cellular senescence.

Another unique function of HGF is its anti-fibrotic action. Early work from our group demonstrated that up-regulation of HGF resulted in a significant decrease in fibrotic area following acute myocardial infarction [54]. Additionally, HGF significantly attenuates endothelial-mesenchymal transition, which is considered to be involved in perivascular fibrosis in the heart [55] and kidney [38]. These anti-fibrotic actions would serve to minimize impediments to tissue regeneration. Intra-muscular fibrosis may inhibit resident stem/progenitor cell migration. Similarly, perivascular fibrosis would limit engraftment of circulating stem/progenitor cells and oxygen diffusion leading to impairment of tissue regeneration and oxygenation. Since the local HGF expression level is decreased in the ischemic hind-limb, where vascular and intramuscular fibrosis cause disease progression, addition of HGF by gene therapy would have a significant impact on neovascularization and regeneration of the organ. On the other hand, anti-VEGF and anti-FGF therapy have been reported to attenuate lung fibrosis by inhibiting inflammation [56, 57]. Taken together, these different properties of HGF, VEGF and bFGF might affect the efficiency of therapeutic angiogenesis (Figure 1).

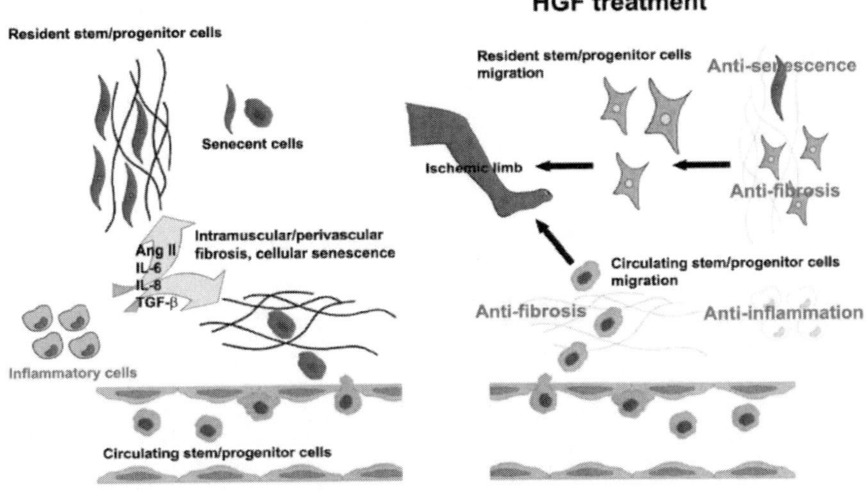


Figure 1. Proposed mechanism of HGF on stem/progenitor cells. Risk factors for CVDs cause inflammation and senescence of vascular progenitor cells. Moreover, tissue fibrosis prevent vascular stem/progenitor function (Left panel). HGF prevents inflammation, cellular senescence, and tissue fibrosis, which enhances vascular stem/progenitor function leading to angiogenesis even in the presence of CVDs risk factors.

## Future Perspective

CLI causes millions of deaths worldwide. Despite a deep desire to apply gene therapy in CLI patients to achieve better quality of life, the dramatic efficacy of angiogenic growth factor therapy that was shown in several animal studies was not fully translated into clinical practice. Comprehensive understanding of the basic biology of neovascularization in pathological conditions in the presence of cardiovascular risk factors will provide important information on the ideal features of future angiogenesis therapy using angiogenic growth factors. Among several angiogenic growth factors, HGF has a unique action on inflammation, cellular senescence, and fibrosis, which are involved in the development of atherosclerosis and subsequent ischemic events. Although it is challenging to translate basic research to the clinical situation, this unique potential of HGF might be more ideal for the treatment of CLI patients with atherosclerosis of the vasculature.

## Acknowledgment

We thank the members of the Department of Clinical Gene Therapy, Osaka University Graduate School of Medicine, for much helpful discussion and technical support.

## Sources of Funding

This work was partially supported by a grant-in-aid from the Organization for Pharmaceutical Safety and Research, a grant-in-aid from the Ministry of Public Health and Welfare, a grant-in-aid from Japan Promotion of Science, and funds from the Ministry of Education, Culture, Sports, Science and Technology, of the Japanese Government.

## Disclosures

R. Morishita received honoraria, consulting fees, and funds from Novartis, Takeda, Shionogi, Asteras, Boehringer Ingelfim, Daiichi-Sankyo, and Pfizer.

# References


[1]     Global status report on noncommunicable diseases 2010. Geneva, World Health Organization, 2011.

[2]     Global atlas on cardiovascular disease prevention and control. Geneva, World Health Organization, 2011.

[3]     Minamino, T; Komuro, I. Vascular cell senescence: contribution to atherosclerosis. *Circ Res.*, 2007 Jan 5, 100(1), 15-26.

[4]     Vasa, M; Breitschopf, K; Zeiher, AM; Dimmeler, S. Nitric oxide activates telomerase and delays endothelial cell senescence. *Circ Res.*, 2000 Sep 29, 87(7), 540-2.

[5]     Maier, JA; Statuto, M; Ragnotti, G. Senescence stimulates U937-endothelial cell interactions. *Exp Cell Res.*, 1993 Sep, 208(1), 270-4.

[6]     Hill, JM; Zalos, G; Halcox, JP; Schenke, WH; Waclawiw, MA; Quyyumi, AA; Finkel, T. Circulating endothelial progenitor cells, vascular function, and cardiovascular risk. *N Engl J Med.*, 2003 Feb 13, 348(7), 593-600.

[7]     Marín, J. Age-related changes in vascular responses: a review. *Mech Ageing Dev.*, 1995 Apr 14, 79(2-3), 71-114. Review.

[8]     Crass, MF 3rd; Borst, SE; Scarpace, PJ. Beta-adrenergic responsiveness in cultured aorta smooth muscle cells. Effects of subculture and aging. *Biochem Pharmacol.*, 1992 Apr 15, 43(8), 1811-5.

[9]     Powell, RJ; Simons, M; Mendelsohn, FO; Daniel, G; Henry, TD; Koga, M; Morishita, R; Annex, BH. Results of a double-blind, placebo-controlled study to assess the safety of intramuscular injection of hepatocyte growth factor plasmid to improve limb perfusion in patients with critical limb ischemia. *Circulation.*, 2008, 118, 58–65.

[10]    Asahara, T; Murohara, T; Sullivan, A; Silver, M; van der Zee, R; Li, T; Witzenbichler, B; Schatteman, G; Isner, JM. Isolation of putative progenitor endothelial cells for angiogenesis. *Science.*, 1997 Feb 14, 275(5302), 964-7.

[11]    Hu, Y; Davison, F; Zhang, Z; Xu, Q. Endothelial replacement and angiogenesis in arteriosclerotic lesions of allografts are contributed by circulating progenitor cells. *Circulation.*, 2003 Dec 23, 108(25), 3122-7. Epub 2003 Dec 1. Erratum in: Circulation. 2009 Sep, 120(10), e82.

[12]    Vasa, M; Fichtlscherer, S; Aicher, A; Adler, K; Urbich, C; Martin, H; Zeiher, AM; Dimmeler, S. Number and migratory activity of circulating endothelial progenitor cells inversely correlate with risk factors for coronary artery disease. *Circ Res.*, 2001 Jul 6, 89(1), E1-7.

[13] Fenton, M; Barker, S; Kurz, DJ; Erusalimsky, JD. Cellular senescence after single and repeated balloon catheter denudations of rabbit carotid arteries. *Arterioscler Thromb Vasc Biol.*, 2001 Feb, 21(2), 220-6.

[14] Minamino, T; Miyauchi, H; Yoshida, T; Ishida, Y; Yoshida, H; Komuro, I. Endothelial cell senescence in human atherosclerosis: role of telomere in endothelial dysfunction. *Circulation.*, 2002 Apr 2, 105(13), 1541-4.

[15] Garfinkel, S; Brown, S; Wessendorf, JH; Maciag, T. Post-transcriptional regulation of interleukin 1 alpha in various strains of young and senescent human umbilical vein endothelial cells. *Proc Natl Acad Sci U S A.*, 1994 Feb 15, 91(4), 1559-63.

[16] Nakajima, M; Hashimoto, M; Wang, F; Yamanaga, K; Nakamura, N; Uchida, T; Yamanouchi, K. Aging decreases the production of PGI2 in rat aortic endothelial cells. *Exp Gerontol.*, 1997 Nov-Dec, 32(6), 685-93.

[17] Mühlhauser, J; Merrill, MJ; Pili, R; Maeda, H; Bacic, M; Bewig, B; Passaniti, A; Edwards, NA; Crystal, RG; Capogrossi, MC. VEGF165 expressed by a replication-deficient recombinant adenovirus vector induces angiogenesis in vivo. *Circ Res.*, 1995 Dec, 77(6), 1077-86.

[18] Gowdak, LH; Poliakova, L; Wang, X; Kovesdi, I; Fishbein, KW; Zacheo, A; Palumbo, R; Straino, S; Emanueli, C; Marrocco-Trischitta, M; Lakatta, EG; Anversa, P; Spencer, RG; Talan, M; Capogrossi, MC. Adenovirus-mediated VEGF(121) gene transfer stimulates angiogenesis in normoperfused skeletal muscle and preserves tissue perfusion after induction of ischemia. *Circulation.*, 2000 Aug 1, 102(5), 565-71.

[19] Takahashi, H; Shibuya, M. The vascular endothelial growth factor (VEGF)/VEGF receptor system and its role under physiological and pathological conditions. *Clin Sci* (Lond)., 2005 Sep, 109(3), 227-41.

[20] Takeshita, S; Zheng, LP; Brogi, E; Kearney, M; Q Pu, L; Bunting, S; Ferrara, N; Symes, JF; Isner, JM. Therapeutic angiogenesis. A single intra arterial bolus of vascular endothelial growth factor augments revascularization in a rabbit ischemic hind limb model. *J Clin Invest.*, 1994 February, 93(2), 662–670.

[21] Walder, CE; Errett, CJ; Bunting, S; Lindquist, P; Ogez, JR; Heinsohn, HG; Ferrara, N; Thomas, GR. Vascular endothelial growth factor augments muscle blood flow and function in a rabbit model of chronic hindlimb ischemia. *J Cardiovasc Pharmacol.*, 1996 Jan, 27(1), 91-8.

[22] Kalka, C; Tehrani, H; Laudenberg, B; Vale, PR; Isner, JM; Asahara, T; Symes, JF. VEGF gene transfer mobilizes endothelial progenitor cells in patients with inoperable coronary disease. *Ann Thorac Surg.*, 2000 Sep, 70(3), 829-34.

[23] Asahara, T; Takahashi, T; Masuda, H; Kalka, C; Chen, D; Iwaguro, H; Inai, Y; Silver, M; Isner, JM. VEGF contributes to postnatal neovascularization by mobilizing bone marrow-derived endothelial progenitor cells. *EMBO J.* 1999 Jul 15, 18(14), 3964-72.

[24] Baumgartner, I; Pieczek, A; Manor, O; Blair, R; Kearney, M; Walsh, K; Isner, JM. Constitutive expression of phVEGF165 after intramuscular gene transfer promotes collateral vessel development in patients with critical limb ischemia. *Circulation.*, 1998 Mar 31, 97(12), 1114-23.

[25] Mäkinen, K; Manninen, H; Hedman, M; Matsi, P; Mussalo, H; Alhava, E; Ylä-Herttuala, S. Increased vascularity detected by digital subtraction angiography after VEGF gene transfer to human lower limb artery: a randomized, placebo-controlled, double-blinded phase II study. *Mol Ther.*, 2002 Jul, 6(1), 127-33.

[26] Rajagopalan, S; Mohler, E 3rd; Lederman, RJ; Saucedo, J; Mendelsohn, FO; Olin, J; Blebea, J; Goldman, C; Trachtenberg, JD; Pressler, M; Rasmussen, H; Annex, BH; Hirsch, AT. Regional Angiogenesis With Vascular Endothelial Growth Factor trial. Regional Angiogenesis with Vascular Endothelial Growth Factor (VEGF) in peripheral arterial disease: Design of the RAVE trial. *Am Heart J.*, 2003 Jun, 145(6), 1114-8.

[27] Ehrbar, M; Djonov, VG; Schnell, C; Tschanz, SA; Martiny-Baron, G; Schenk, U; Wood, J; Burri, PH; Hubbell, JA; Zisch, AH. Cell-demanded liberation of VEGF121 from fibrin implants induces local and controlled blood vessel growth. *Circ Res.*, 2004 Apr30, 94(8), 1124-32.

[28] Burger, PE; Coetzee, S; McKeehan, WL; Kan, M; Cook, P; Fan, Y; Suda, T; Hebbel, RP; Novitzky, N; Muller, WA; Wilson, EL. Fibroblast growth factor receptor-1 is expressed by endothelial progenitor cells. *Blood.*, 2002 Nov 15, 100(10), 3527-35.

[29] Gonçalves, LM. Fibroblast growth factor-mediated angiogenesis for the treatment of ischemia. Lessons learned from experimental models and early human experience. *Rev Port Cardiol.*, 1998 Oct, 17 Suppl 2, II11-20.

[30] Comerota, AJ; Throm, RC; Miller, KA; Henry, T; Chronos, N; Laird, J; Sequeira, R; Kent, CK; Bacchetta, M; Goldman, C; Salenius, JP; Schmieder, FA; Pilsudski, R. Naked plasmid DNA encoding fibroblast growth factor type 1 for the treatment of end-stage unreconstructible lower extremity ischemia: preliminary results of a phase I trial. *J Vasc Surg.*, 2002 May, 35(5), 930-6.

[31]  Nikol, S; Baumgartner, I; Van Belle, E; Diehm, C; Visoná, A; Capogrossi, MC; Ferreira-Maldent, N; Gallino, A; Wyatt, MG; Wijesinghe, LD; Fusari, M; Stephan, D; Emmerich, J; Pompilio, G; Vermassen, F; Pham, E; Grek, V; Coleman, M; Meyer, F. TALISMAN 201 investigators. Therapeutic angiogenesis with intramuscular NV1FGF improves amputation-free survival in patients with critical limb ischemia. *Mol Ther.*, 2008 May, 16(5), 972-8.

[32]  Belch, J; Hiatt, WR; Baumgartner, I; Driver, IV; Nikol, S; Norgren, L; Van Belle, E. TAMARIS Committees and Investigators. Effect of fibroblast growth factor NV1FGF on amputation and death: a randomised placebo-controlled trial of gene therapy in critical limb ischaemia. *Lancet.*, 2011 Jun 4, 377(9781), 1929-37.

[33]  Wojakowski, W; Tendera, M; Michałowska, A; Majka, M; Kucia, M; Maślankiewicz, K; Wyderka, R; Ochała, A; Ratajczak, MZ. Mobilization of CD34/CXCR4+, CD34/CD117+, c-met+ stem cells, and mononuclear cells expressing early cardiac, muscle, and endothelial markers into peripheral blood in patients with acute myocardial infarction. *Circulation.*, 2004 Nov 16, 110(20), 3213-20.

[34]  Powell, RJ; Goodney, P; Mendelsohn, FO; Moen, EK; Annex, BH. HGF-0205 Trial Investigators. Safety and efficacy of patient specific intramuscular injection of HGF plasmid gene therapy on limb perfusion and wound healing in patients with ischemic lower extremity ulceration: results of the HGF-0205 trial. *J Vasc Surg.*, 2010 Dec, 52(6), 1525-30.

[35]  Shigematsu, H; Yasuda, K; Iwai, T; Sasajima, T; Ishimaru, S; Ohashi, Y; Yamaguchi, T; Ogihara, T; Morishita, R. Randomized, double-blind, placebo-controlled clinical trial of hepatocyte growth factor plasmid for critical limb ischemia. *Gene Ther.*, 2010 Sep, 17(9), 1152-61.

[36]  Makino, H; Aoki, M; Hashiya, N; Yamasaki, K; Azuma, J; Sawa, Y; Kaneda, Y; Ogihara, T; Morishita, R. Long-term follow-up evaluation of results from clinical trial using hepatocyte growth factor gene to treat severe peripheral arterial disease. *Arterioscler Thromb Vasc Biol.*, 2012 Oct, 32(10), 2503-9.

[37]  Taniyama, Y; Morishita, R; Aoki, M; Hiraoka, K; Yamasaki, K; Hashiya, N; Matsumoto, K; Nakamura, T; Kaneda, Y; Ogihara, T. Angiogenesis and antifibrotic action by hepatocyte growth factor in cardiomyopathy. *Hypertension.*, 2002 Jul, 40(1), 47-53.

[38]  Iekushi, K; Taniyama, Y; Kusunoki, H; Azuma, J; Sanada, F; Okayama, K; Koibuchi, N; Iwabayashi, M; Rakugi, H; Morishita, R. Hepatocyte growth factor attenuates transforming growth factor-β-angiotensin II

crosstalk through inhibition of the PTEN/Akt pathway. *Hypertension.*, 2011 Aug, 58(2), 190-6.

[39]  Ueki, T; Kaneda, Y; Tsutsui, H; Nakanishi, K; Sawa, Y; Morishita, R; Matsumoto, K; Nakamura, T; Takahashi, H; Okamoto, E; Fujimoto, J. Hepatocyte growth factor gene therapy of liver cirrhosis in rats. *Nat Med.*, 1999 Feb, 5(2), 226-30.

[40]  Sanada, F; Taniyama, Y; Iekushi, K; Azuma, J; Okayama, K; Kusunoki, H; Koibuchi, N; Doi, T; Aizawa, Y; Morishita, R. Negative action of hepatocyte growth factor/c-Met system on angiotensin II signaling via ligand-dependent epithelial growth factor receptor degradation mechanism in vascular smooth muscle cells. *Circ Res.*, 2009 Sep 25, 105(7), 667-75,

[41]  Sanada, F; Taniyama, Y; Azuma, J; Iekushi, K; Dosaka, N; Yokoi, T; Koibuchi, N; Kusunoki, H; Aizawa, Y; Morishita, R. Hepatocyte growth factor, but not vascular endothelial growth factor, attenuates angiotensin II-induced endothelial progenitor cell senescence. *Hypertension.*, 2009 Jan, 53(1), 77-82.

[42]  Vlassara, H; Cai, W; Chen, X; Serrano, EJ; Shobha, MS; Uribarri, J; Woodward, M; Striker, GE. Managing chronic inflammation in the aging diabetic patient with CKD by diet or sevelamer carbonate: a modern paradigm shift. *J Gerontol A Biol Sci Med Sci.*, 2012 Dec, 67(12), 1410-6. doi: 10.1093/gerona/gls195. Epub 2012 Oct 29.

[43]  Hansson, GK. Inflammation, atherosclerosis, and coronary artery disease. *N Engl J Med.*, 2005 Apr 21, 352(16), 1685-95.

[44]  Mantzouranis, EC; Rosen, FS; Colten, HR. Reticuloendothelial clearance in cystic fibrosis and other inflammatory lung diseases. *N Engl J Med.*, 1988 Aug 11, 319(6), 338-43.

[45]  Kuilman, T; Michaloglou, C; Vredeveld, LC; Douma, S; van Doorn, R; Desmet, CJ; Aarden, LA; Mooi, WJ; Peeper, DS. Oncogene-induced senescence relayed by an interleukin-dependent inflammatory network. *Cell.*, 2008 Jun 13, 133(6), 1019-31.

[46]  Chandeck, C; Mooi, WJ. Oncogene-induced cellular senescence. *Adv Anat Pathol.* 2010 Jan, 17(1), 42-8.

[47]  Imanishi, T; Tsujioka, H; Akasaka, T. Endothelial progenitor cell senescence--is there a role for estrogen? *Ther Adv Cardiovasc Dis.*, 2010 Feb, 4(1), 55-69.

[48]  Shimizu, K; Taniyama, Y; Sanada, F; Azuma, J; Iwabayashi, M; Iekushi, K; Rakugi, H; Morishita, R. Hepatocyte growth factor inhibits lipopolysaccharide-induced oxidative stress via epithelial growth factor

receptor degradation. *Arterioscler Thromb Vasc Biol.*, 2012 Nov, 32(11), 2687-93.

[49]  Kaga, T; Kawano, H; Sakaguchi, M; Nakazawa, T; Taniyama, Y; Morishita, R. Hepatocyte growth factor stimulated angiogenesis without inflammation: differential actions between hepatocyte growth factor, vascular endothelial growth factor and basic fibroblast growth factor. *Vascul Pharmacol.*, 2012 Aug 19, 57(1), 3-9.

[50]  Ohtani, K; Egashira, K; Hiasa, K; Zhao, Q; Kitamoto, S; Ishibashi, M; Usui, M; Inoue, S; Yonemitsu, Y; Sueishi, K; Sata, M; Shibuya, M; Sunagawa, K. Blockade of vascular endothelial growth factor suppresses experimental restenosis after intraluminal injury by inhibiting recruitment of monocyte lineage cells. *Circulation.*, 2004 Oct 19, 110(16), 2444-52.

[51]  Van Belle, E; Witzenbichler, B; Chen, D; Silver, M; Chang, L; Schwall, R; Isner, JM. Potentiated angiogenic effect of scatter factor/hepatocyte growth factor via induction of vascular endothelial growth factor: the case for paracrine amplification of angiogenesis. *Circulation.*, 1998 Feb 3, 97(4), 381-90.

[52]  Xin, X; Yang, S; Ingle, G; Zlot, C; Rangell, L; Kowalski, J; Schwall, R; Ferrara, N; Gerritsen, ME. Hepatocyte growth factor enhances vascular endothelial growth factor-induced angiogenesis in vitro and in vivo. *Am J Pathol.*, 2001Mar, 158(3), 1111-20.

[53]  Min, JK; Lee, YM; Kim, JH; Kim, YM; Kim, SW; Lee, SY; Gho, YS; Oh, GT; Kwon, YG. Hepatocyte growth factor suppresses vascular endothelial growth factor-induced expression of endothelial ICAM-1 and VCAM-1 by inhibiting the nuclear factor-kappaB pathway. *Circ Res.*, 2005 Feb 18, 96(3), 300-7.

[54]  Taniyama, Y; Morishita, R; Nakagami, H; Moriguchi, A; Sakonjo, H; Shokei-Kim; Matsumoto, K; Nakamura, T; Higaki, J; Ogihara, T. Potential contribution of a novel antifibrotic factor, hepatocyte growth factor, to prevention of myocardial fibrosis by angiotensin II blockade in cardiomyopathic hamsters. *Circulation.*, 2000 Jul 11, 102(2), 246-52.

[55]  Okayama, K; Azuma, J; Dosaka, N; Iekushi, K; Sanada, F; Kusunoki, H; Iwabayashi, M; Rakugi, H; Taniyama, Y; Morishita, R. Hepatocyte growth factor reduces cardiac fibrosis by inhibiting endothelial-mesenchymal transition. *Hypertension.*, 2012 May, 59(5), 958-65.

[56]  Hamada, N; Kuwano, K; Yamada, M; Hagimoto, N; Hiasa, K; Egashira, K; Nakashima, N; Maeyama, T; Yoshimi, M; Nakanishi, Y. Anti-

vascular endothelial growth factor gene therapy attenuates lung injury and fibrosis in mice. *J Immunol.*, 2005 Jul 15, 175(2), 1224-31.

[57] Chaudhary, NI; Roth, GJ; Hilberg, F; Müller-Quernheim, J; Prasse, A; Zissel, G; Schnapp, A; Park, JE. Inhibition of PDGF, VEGF and FGF signalling attenuates fibrosis. *Eur Respir J.*, 2007 May, 29(5), 976-85. Epub 2007 Feb 14.

[58] Kusumanto, YH; van Weel, V; Mulder, NH; Smit, AJ; van den Dungen, JJ; Hooymans, JM; Sluiter, WJ; Tio, RA; Quax, PH; Gans, RO; Dullaart, RP; Hospers, GA. Treatment with intramuscular vascular endothelial growth factor gene compared with placebo for patients with diabetes mellitus and critical limb ischemia: a double-blind randomized trial. *Hum Gene Ther.*, 2006, 17:683– 691.

[59] Morishita, R; Aoki, M; Hashiya, N; Makino, H; Yamasaki, K; Azuma, J; Sawa, Y; Matsuda, H; Kaneda, Y; Ogihara, T. Safety evaluation of clinical gene therapy using hepatocyte growth factor to treat peripheral arterial disease. *Hypertension.*, 2004 Aug, 44(2), 203-9.

In: Progenitor Cells                    ISBN: 978-1-62808-994-3
Editors: P. M. Horton, B. E. Lawrence  © 2013 Nova Science Publishers, Inc.

*Chapter 6*

# Role of MicroRNAs in Endothelial Progenitor Cells: Implication for Cardiac Repair

*E. Goretti[1], D. R. Wagner[1,2] and Y. Devaux[1,*]*
[1]Laboratory of Cardiovascular Research, Public Research Centre - Health, Luxembourg, Luxembourg
[2]Division of Cardiology, Centre Hospitalier, Luxembourg, Luxembourg

## Abstract

Endothelial progenitor cells (EPC) are mobilized after myocardial infarction (MI) from the bone marrow to injured sites of the heart where they participate in cardiac repair by revascularization of ischemic tissues. Endothelial progenitor cells have been actively studied, but their exact phenotype and regenerative properties are still controversial. Small trials with progenitor cells of different origins showed modest clinical benefits. It is assumed that a better understanding of the biology of EPC will contribute to improve their therapeutic potential. MicroRNAs (miRNAs) are small single-stranded non-coding RNAs that modulate gene

---

[*] Correspondence to Yvan Devaux, Laboratory of Cardiovascular Research, Centre de Recherche Public de la Santé (CRP-Santé), 84 Val Fleuri L-1526, Luxembourg, Luxembourg, Tel +352 26970300; Fax +352 26970396; Email: yvan.devaux@crp-sante.lu.

expression by interacting post transcriptionally with protein-coding RNAs. MicroRNAs regulate multiple biological processes involved in cardiac development and disease. While many studies addressed the role of miRNAs in cardiac cells, less is known of the effect of miRNAs in EPC. Recent studies showed that miRNAs indeed regulate the biology of EPC. Since novel technologies to enhance or blunt the functions of miRNAs have been recently developed, it is conceivable that miRNAs may become promising new therapeutic tools. This article will review the recent advances in the knowledge of the effects of miRNAs in EPC and will discuss how miRNAs could be manipulated to improve the regenerative capacities of EPC in the diseased heart.

**Keywords:** MicroRNAs; endothelial progenitor cell; cardiovascular diseases; cardiac repair; angiogenesis

# 1. Introduction

MicroRNAs (miRNAs) are small single stranded and non-coding RNAs (about 18-22 nucleotides) discovered firstly in 1993 in Caenorhabditis elegans [1]. They have the ability to inhibit gene expression by targeting messenger RNAs (mRNAs). The biogenesis of miRNAs is complex and implies the evolution of a primary transcript called pri-miRNA to a mature miRNA. Once mature, this miRNA can interact with its target mRNA by base pairing. If both miRNA and mRNA sequences perfectly match, mRNA degradation is induced. If not, mRNA translation is repressed [2]. Today, more than 25000 miRNAs have been identified throughout 193 species (http://www.mirbase. org, version 19). In humans, a little more than 2000 mature miRNAs have been discovered, and this number continues to grow. Some miRNAs display a certain level of tissue or cell specificity, which is the case for the heart.

MicroRNAs can target more than half of the human genome, and multiple mRNAs can be regulated by one unique miRNA. Thus, miRNAs can regulate many biological pathways or phenomena involved in cardiac development and cardiovascular diseases (CVDs), such as hypertrophy and angiogenesis.

Among CVDs, myocardial infarction (MI) is one of the leading causes of death worldwide. This widespread manifestation of coronary artery disease (CAD) triggers the participation of diverse cell population, among which endothelial progenitor cells (EPC). These cells, discovered in 1997 [3], are recruited from the bone marrow to the infarct zone where they participate in cardiac repair by increasing revascularization. They share both endothelial and progenitor functions, properties and markers. Despite many studies performed in these cells, their phenotype and biology are still controversial. This is principally due to the multiplicity of techniques used for their isolation and culture, but also to their low abundance in the circulation [4]. However, two major types of EPC have been identified and characterized: early EPC which secrete pro-angiogenic factors and late EPC, which proliferate and form spontaneously a microtubular network.

Helping revascularization after MI by targeting EPC appears to be a promising strategy [5], but EPC are difficult to recruit and engraft in the heart [6]. However, a better characterization of the phenotype and properties of EPC may help to improve recruitment and engraftment in the heart.

MicroRNAs are expressed in progenitor cells and could help to understand some properties and better characterize the phenotype of EPC. Since recent technologies allow for the modulation of their expression, either by inhibitors (antimiRs) or mimics (premiRs) [7], miRNAs constitute a family of promising tools to improve the regenerative capacities of EPC.

# 2. MicroRNAs

## 2.1. MicroRNAs Biogenesis and Mode of Action

MicroRNAs are small (18-22 nucleotides), single-stranded and non-coding RNAs that inhibit gene expression by interaction with mRNA [2].

The biogenesis of miRNAs and their mode of action are complex and still unclear. It begins with the transcription of a long precursor capped and polyadenylated miRNA by RNA polymerase II, called pri-miRNA (Figure 1). This pri-miRNA is supported by the RNase III drosha associated with other cofactors to form a pre-miRNA.

The pre-miRNA is about 65-70 nucleotides long, and hairpin shaped. It is exported to the cytoplasm by exportin-5 where another RNase II, Dicer, cuts its hairpin: a duplex miRNA-miRNA* is formed, about 18-24 nucleotides long. One strand is degraded whereas the other one is retained into the complex. This latter strand will become the mature miRNA. This mature miRNAs will then be loaded into another complex, RISC (RNA-induced silencing complex), which will stabilize the interaction between the mature miRNA and its mRNA target. The miRNA contains a "seed" sequence: nucleotides 2-7 of its 5' untranslated region (UTR). This sequence interacts by base pairing with the 3'UTR of the mRNA target. If the base pairing contains mismatches, only translational repression of the mRNA will occur. If the base pairing is perfect, cleavage and degradation of the miRNA-mRNA duplex will be induced. Both pathways lead to inhibition of the target gene expression.

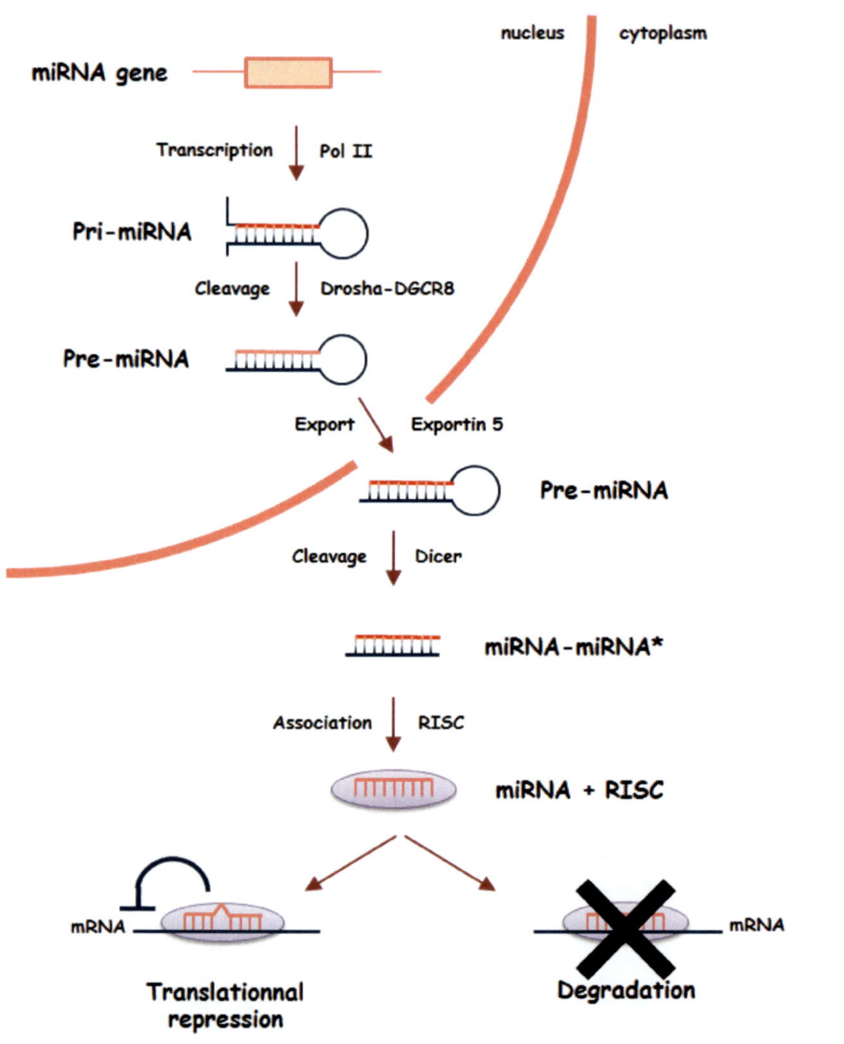


Figure 1. miRNAs biogenesis.

It is important to note that a unique miRNA can target multiple mRNAs (more than 200) and that one single mRNA can be targeted by several miRNAs. As such, miRNAs can regulate more than 30% of all protein coding genes. This shows the degree of the regulation of gene expression by miRNAs.

The presence of miRNAs was discovered in biological fluids, such as blood [8]. In body fluids, miRNAs are very stable and resistant to RNase degradation.

This discovery suggested that miRNAs could be used as biomarkers. Indeed, their presence in blood can originate from dead cells (passive release) or from living cells (active secretion). Circulating miRNAs can thus act as paracrine factors, after inclusion into diverse components, such as microvesicles, exosomes, apoptotic bodies, or embedded into protein complexes, lipids or high-density lipoproteins. MiRNAs have been shown to be secreted by different cell types and regulate gene expression in other cell types [9, 10].

## 2.2. MicroRNAs in Cardiac Function

MicroRNAs are implicated in many biological mechanisms and pathways involved in cardiac function such as development, apoptosis, differentiation, proliferation. Initial studies have demonstrated that circulating levels of miRNAs expression are regulated by cardiac stress, such as myocardial infarction [11, 12], suggesting that they may be useful biomarkers in this setting. However, more recent large scale studies tempered their use as diagnostic biomarkers of MI [13, 14].

## 2.3. MicroRNAs as Therapeutic Tools

Modulating the expression and function of miRNAs appears as a promising strategy to treat CVDs. MicroRNAs expression can be increased by addition of pre-miRs or mimicry, and can be decreased by antimiRs or antagomiRs [7]. These agents can be used both for in vitro and in vivo experiments. The most developed method for the increase of a miRNA expression is to deliver the mimicry associated with adeno-associated viruses [15]. For antimiR therapy, several possibilities with specific pharmacokinetics and pharmacodynamics exist, such as antimiRs with locked nucleic acid modification. Targeting an entire miR-family (by targeting the seed sequence shared by the member of this family) was showed to be feasible [16].

One antimiR, a miR-122 inhibitor against hepatitis C, called Miravirsen (Santaris Pharma), is currently in phase II trials [17].

# 3. Endothelial Progenitor Cells

## 3.1. Characteristics and Properties

Endothelial progenitor cells were discovered in human blood by Asahara et al. in 1997 [3].

Isolated from peripheral blood (and rarely from cord blood), EPC share both endothelial and progenitor cells properties. They were first characterized by their ability to bind lectins and to take up acetylated low density lipoprotein and also by the expression at the cell surface of the progenitor markers CD133 and CD34 and the endothelial marker vascular endothelial growth factor receptor 2 (VEGFR2). EPC have been actively studied over the past fifteen years, however their phenotype and functional properties are still controversial [4]. This is mostly due to the number of isolation procedures and culture methods used and to their low frequency in the circulation. Thus, different populations with different properties and cell surface markers have been obtained. However, two main populations of EPC were extensively studied: early EPC and late EPC [18]. Early EPC are specialized in the secretion of pro-angiogenic factors. These cells, unable to proliferate and to form microtubular structure in matrigel assay, have the capacity to integrate a pre-formed network of endothelial cells. Early EPC are quite heterogeneous and share leukocyte features. Indeed, they express: the progenitor cell markers CD34 and CD133, the pan-leukocyte marker CD45, the endothelial cell markers CD31 and von Willebrand Factor (vWF), and the monocyte/macrophage marker CD14 [18]. On the other hand, late EPC are able to form microtubule networks [18]. Their capacity to secrete cytokines is minor compared to early EPC, but they have proliferation capacity. Late EPC are more homogenous morphologically and are close to endothelial cells compared to early EPC. They express the progenitor cell markers CD34 and CD133, and several endothelial markers such as CD31, CD141, CD105, CD146, CD144, vWF, flk-1. Early and late EPC have complementary functions which allow stimulating revascularization [19]. Indeed, early EPC secrete pro-angiogenic factors that stimulate late EPC to form microtubules.

## 3.2. Endothelial Progenitor Cells and Cardiovascular Disease

After MI, EPC are recruited from bone marrow to the infarct zone in order to repair the ischemic heart by increasing revascularization and thus blood supply [5]. Clinical trials showed that autologous transplantation of bone marrow-derived cell is safe, feasible, and potentially beneficial for patients with peripheral arterial disease [20]. However, transplantation of EPC showed only modest benefits due to the poor engraftment of cells at sites of injury [6].

Different strategies are currently under investigation to amplify their number, recruitment and retention in the heart. One of these strategies is to stimulate their recruitment by interacting with the receptor-ligand couple regulating their migration, CXCR4/ stromal cell-derived factor 1α (SDF1α). Indeed, increased CXCR4 expression amplifies the migration of EPC [21] which ultimately results in stimulation of tissue repair and capillary density [22].

However, this strategy lacks effectiveness and novel approaches are needed to enhance the regenerative capacities of EPC. As recently suggested [23], miRNAs seem to be attractive targets to achieve this goal.

# 4. MicroRNAs and EPC

This chapter summarizes current knowledge of the function of miRNAs in EPC (figure 2) and discusses how these functions can be targeted to stimulate cardiac repair.

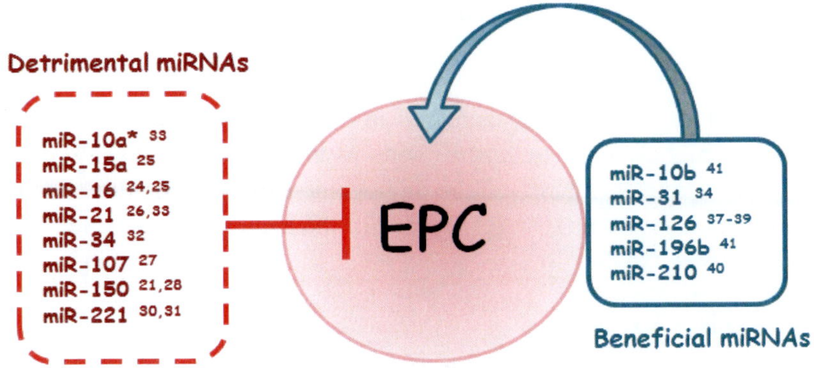


Figure 2. Functional miRNAs in EPC.

## 4.1. Detrimental miRNAs

### 4.1.1. MiR-15a/-16

Using microarrays, we profiled miRNA expression of both early and late EPC isolated from peripheral blood of healthy volunteers [24]. We observed that several miRNAs were differentially expressed between these two cell types. Among these, we noticed that early EPC overexpressed five members of the miR-16 family, compared to late EPC. Three members of the cyclins family of cell-cycle regulating proteins were predicted as target genes of miR-16 family members. These 3 genes were down-regulated in early EPC, consistently with the overexpression of miR-16 family members, and with the lack of proliferation potential of these cells. Inhibition of miR-16 by antimiRs allowed early EPC to re-enter the cell cycle, to differentiate towards a more endothelial phenotype, and to secrete the pro angiogenic cytokine IL-8. These data suggested that antagonism of miR-16 may improve the healing capacities of early EPC. MiR-15 (a member of miR-16 family) and miR-16 were also found to be overexpressed in circulating angiogenic cells (early EPC) isolated from patients with critical limb ischemia compared to controls [25]. An increase of these miRNAs in EPC from controls inhibited migration and survival, whereas a decrease in EPC from patients with critical limb ischemia stimulated migration. EPC from healthy donors transfected with antimiR-15a/-16 and transplanted into mice with critical limb ischemia increased angiogenesis, induced blood flow recovery and enhanced muscular arteriole density. In addition, circulating levels of miR-15a/-16 were higher in patients with critical limb ischemia compared to controls. These results are consistent with our findings [24] and strengthen the concept that inhibiting miR-15a/-16 may be a promising strategy to stimulate the angiogenic capacities of EPC.

### 4.1.2. MiR-21

MiR-21 expression was increased in EPC from patients with CAD, and was even higher after treatment with asymmetric dimethylarginine [26]. Inhibition of miR-21 increased the angiogenic capacities of EPC, identifying this miRNA as a potential target to enhance the angiogenic properties of EPCs, in patients with CAD.

### 4.1.3. MiR-107

Meng et al. reported that miR-107 is overexpressed in EPC during hypoxia, and this prevented endothelial differentiation [27]. Inhibition of miR-107 relieved this blockade and promoted differentiation. This effect involved hypoxia inducible factor 1β. These data are relevant to the MI context, in which cells are subjected to oxygen deprivation.

### 4.1.4. MiR-150

In early EPC, we observed a high level of miR-150 expression, associated with a low level of CXCR4 [21]. Adenosine, a nucleoside with cardioprotective properties, was able to decrease miR-150 expression, thereby increasing CXCR4 expression and EPC migration. This result is consistent with the study by Tano et al. showing that, during ischemic stress, miR-150 down-regulates CXCR4 in bone-marrow derived mononuclear cells and that migration is enhanced when miR-150 is inhibited [28]. We also observed that patients with low levels of miR-150 after acute MI develop more left ventricular remodeling [29].

### 4.1.5. MiR-221/-222

Minami et al. observed that miR-221 and miR-222 were overexpressed by early EPC isolated from CAD patients and were inversely correlated with the number of EPC in these patients [30]. Interestingly, expression level of these 2 miRNAs was decreased after 12 months of treatment with atorvastatin, and the number of EPC was enhanced. This suggested that miRNAs may be modulated by atorvastatin. MiR-221 was also increased in EPC isolated from patients with atherosclerosis, as compared to controls, and overexpression led to a decrease of the proliferation of EPC [31].

### 4.1.6. Senescence Related miRNAs : miR-34a/-10a*/-21

Several miRNAs involved in cell senescence are expressed in EPC. MiR-34a, known as tumor suppressor, induces EPC senescence and inhibits their angiogenic properties [32]. MiR-10a* and miR-21 are overexpressed by aged EPC and their inhibition decreases senescence, enhances self-renewal and angiogenic potential. On the contrary, their overexpression in young EPC induces senescence, decreases self-renewal and reduces angiogenesis [33].

## 4.2. Beneficial miRNAs

### 4.2.1. MiR-31

Cheng et al. investigated the differences between peripheral blood EPC and cord blood EPC [34]. EPC from cord blood expressed more pro-angiogenic miRNAs, such as miR-31. Inhibition of miR-31 in these cord blood cells blunted their migration and microtubule formation, whereas overexpression of miR-31 in peripheral blood EPC enhanced their angiogenic capacities.

### 4.2.2. MiR-126

The angiomiR-126 is highly expressed in endothelial cells, and has been extensively studied in EPC. In a study comparing miRNAs expressed by EPC from patients with CAD and controls, miR-126 was down-regulated and miR-92a was up-regulated by EPC isolated from patients with CAD [35]. Atorvastatin modulated the expression of these miRNAs by increasing the pool of circulating EPC. In another study, expression of miR-126 and miR-208-5p in EPC was associated with the outcome of patients with ischemic cardiomyopathy and patients with chronic heart failure (CHF) [36], suggesting that these miRNAs may be used as prognostic biomarkers in this setting. MiR-126 and miR-130a were decreased in early EPC and CD34+ cells isolated from CHF patients compared to controls [37], providing a possible explanation for the lack of effect on neovascularization of transplantation of early EPC from CHF patients to nude mice subjected to MI. Inhibition of miR-126 in EPC isolated from healthy subjects decreased their angiogenic properties whereas increase of miR-126 in EPC isolated from CHF patients had beneficial effects. In endothelial cells, miR-126 inhibits SDF1α expression [38]. In coculture experiments, inhibition of miR-126 in human umbilical vein endothelial cells increased SDF1α expression and this resulted in an increase of the migration of CD34+ EPC. Treatment with antimiR-126 of mice subjected to hindlimb ischemia enhanced the migration of Sca-1+/Lin2 progenitor cells. Moreover, inhibition of miR-126 in CD34+ cells decreased their angiogenic properties, and increase of miR-126 had opposite effects [39]. Interestingly, level of miR-126 was elevated in supernatants of CD34+ cells, and this appeared to be consecutive to secretion in microvesicles or exosomes. This suggested that miR-126 is secreted by CD34+ cells, and acts as a paracrine pro-angiogenic factor.

### 4.2.3. MiR-210

Alaiti et al. cultured CD34+ cord blood cells in a medium supplemented with vascular endothelial growth factor (VEGF) [40]. The cells formed a dense network in matrigel assay, and enhanced tissue perfusion in ischemic hindlimb model in mice. These properties were related to miR-210 expression, which was overexpressed by VEGF. Inhibition of miR-210 prevented the pro-angiogenic effect of VEGF. The opposite effect was observed with miR-210 mimic.

### 4.2.4. Cancer-Related miRNAs

Plummer et al. investigated the role of EPC in the angiogenic switch, an important element of tumor progression [41]. They found that genetic inhibition of Dicer led to a decrease of the number of circulating EPC, impairment of angiogenesis, and reduction of tumor growth. Two miRNAs, miR-10b and miR-196b, were identified as potential targets to inhibit tumor angiogenesis. These miRNAs have some potential in a cardiovascular context, and it can be expected that stimulation of their expression could help to amplify the angiogenic capacities of EPC.

## 4.3. MiRNAs in Microvesicles from EPC

Cantaluppi et al. reported that microvesicles secreted by EPC are capable of increasing neoangiogenesis of human pancreatic islets [42]. These microvesicles contained angiomiR-126 and angiomiR-196, and inhibition of these miRNAs inhibited these angiogenic effects. In addition, microvesicles isolated from EPC protect the kidney from ischemic injury, partially through a miRNA-dependent mechanism [43]. In nude mice subjected to hindlimb ischemia, the injection of EPC-derived microvesicles increased limb perfusion and decreased the injury [44]. RNase-inactivated microvesicles and microvesicles derived from DICER-knockdown EPC had impaired effect, suggesting that miRNAs trapped in microvesicles could be the source of the regenerative effect of EPC. These data support the notion that miRNAs secreted into microvesicles participate in the pro-angiogenic paracrine effect of EPC.

# Conclusion

Studies mentioned in this commentary are supportive of the concept that miRNAs are promising tools to improve the function of EPC and thereby to stimulate cardiac repair. This is relevant since clinical trials showed only modest effects of cell transplantation, partially due to the difficulty to select the optimal cell population and the poor retention of cells at sites of injury. The benefit of miRNAs is afforded to their capacity to target multiple protein coding genes, thus having the potential to modulate complex biological processes such as angiogenesis. However, the pleiotropic properties of miRNAs may also lead to adverse side effects.

# Perspectives

Continuation to better characterize the phenotype and functions of EPC, through miRNAs for instance, may allow finding new strategies to improve their therapeutic potential. The role of miRNAs in the different types of EPC will have to be well defined. It would be interesting to investigate the benefit of targeting not only one but clusters of miRNAs. Recent studies have shown promising results in this aspect.

Medication appears to modify miRNAs expression, and this will have to be taken into account in future studies.

Finally, the finding that microvesicles secreted by EPC contain miRNAs, which may act as paracrine factors, is appealing and needs further investigation.

*Funding*: Emeline Goretti and Yvan Devaux are supported by grants from the National Research Fund of Luxembourg.

*No conflict of interest*

# References

[1]    Lee RC, Feinbaum RL, Ambros V. The c. Elegans heterochronic gene lin-4 encodes small rnas with antisense complementarity to lin-14. *Cell,* 1993 75,843-854.

[2]    Bartel DP. Micrornas: Genomics, biogenesis, mechanism, and function. *Cell,* 2004 116,281-297.

[3]    Asahara T, Murohara T, Sullivan A, Silver M, van der Zee R, Li T, Witzenbichler B, Schatteman G, Isner JM. Isolation of putative progenitor endothelial cells for angiogenesis. *Science,* 1997 275,964-967.

[4]    Fadini GP, Losordo D, Dimmeler S. Critical reevaluation of endothelial progenitor cell phenotypes for therapeutic and diagnostic use. *Circ Res,* 2012 110,624-637.

[5]    Siddique A, Shantsila E, Lip GY, Varma C. Endothelial progenitor cells: What use for the cardiologist? *J Angiogenes Res,* 2010 2,6.

[6]    Tongers J, Losordo DW, Landmesser U. Stem and progenitor cell-based therapy in ischaemic heart disease: Promise, uncertainties, and challenges. *Eur Heart J,* 2011 32,1197-1206.

[7]    van Rooij E, Olson EN. Microrna therapeutics for cardiovascular disease: Opportunities and obstacles. Nature reviews. *Drug Discov,* 2012 11,860-872.

[8]    Mitchell PS, Parkin RK, Kroh EM, Fritz BR, Wyman SK, Pogosova-Agadjanyan EL, Peterson A, Noteboom J, O'Briant KC, Allen A, Lin DW, Urban N, Drescher CW, Knudsen BS, Stirewalt DL, Gentleman R, Vessella RL, Nelson PS, Martin DB, Tewari M. Circulating micrornas as stable blood-based markers for cancer detection. *Proc Natl Acad Sci USA,* 2008 105,10513-10518.

[9]    Vickers KC, Palmisano BT, Shoucri BM, Shamburek RD, Remaley AT. Micrornas are transported in plasma and delivered to recipient cells by high-density lipoproteins. *Nature Cell Biol,* 2011 13,423-433.

[10]    Montecalvo A, Larregina AT, Shufesky WJ, Stolz DB, Sullivan ML, Karlsson JM, Baty CJ, Gibson GA, Erdos G, Wang Z, Milosevic J, Tkacheva OA, Divito SJ, Jordan R, Lyons-Weiler J, Watkins SC, Morelli AE. Mechanism of transfer of functional micrornas between mouse dendritic cells via exosomes. *Blood,* 2012 119,756-766.

[11]    D'Alessandra Y, Devanna P, Limana F, Straino S, Di Carlo A, Brambilla PG, Rubino M, Carena MC, Spazzafumo L, De Simone M, Micheli B, Biglioli P, Achilli F, Martelli F, Maggiolini S, Marenzi G, Pompilio G, Capogrossi MC. Circulating micrornas are new and sensitive biomarkers of myocardial infarction. *Eur Heart J,* 2010 31,2765-2773.

[12]    Corsten MF, Dennert R, Jochems S, Kuznetsova T, Devaux Y, Hofstra L, Wagner DR, Staessen JA, Heymans S, Schroen B. Circulating microrna-208b and microrna-499 reflect myocardial damage in cardiovascular disease. *Circ Cardiovasc Genet,* 2010 3,499-506.

[13]    Devaux Y, Vausort M, Goretti E, Nazarov PV, Azuaje F, Gilson G, Corsten MF, Schroen B, Lair ML, Heymans S, Wagner DR. Use of circulating micrornas to diagnose acute myocardial infarction. *Clin Chem,* 2012 58,559-567.

[14] Widera C, Gupta SK, Lorenzen JM, Bang C, Bauersachs J, Bethmann K, Kempf T, Wollert KC, Thum T. Diagnostic and prognostic impact of six circulating micrornas in acute coronary syndrome. *J Mol Cell Cardiol,* 2011 51,872-875.

[15] Eulalio A, Mano M, Dal Ferro M, Zentilin L, Sinagra G, Zacchigna S, Giacca M. Functional screening identifies mirnas inducing cardiac regeneration. *Nature,* 2012 492,376-381.

[16] Bernardo BC, Gao X-M, Winbanks CE, Boey EJH, Tham YK, Kiriazis H, Gregorevic P, Obad S, Kauppinen S, Du X-J, Lin RCY, McMullen JR. Therapeutic inhibition of the mir-34 family attenuates pathological cardiac remodeling and improves heart function. *Proc Natl Acad Sci USA,* 2012 109,17615-17620.

[17] Janssen HL, Reesink HW, Lawitz EJ, Zeuzem S, Rodriguez-Torres M, Patel K, van der Meer AJ, Patick AK, Chen A, Zhou Y, Persson R, King BD, Kauppinen S, Levin AA, Hodges MR. Treatment of hcv infection by targeting microrna. *New Engl J Med,* 2013, March 27.

[18] Hur J, Yoon CH, Kim HS, Choi JH, Kang HJ, Hwang KK, Oh BH, Lee MM, Park YB. Characterization of two types of endothelial progenitor cells and their different contributions to neovasculogenesis. *Arterioscler Thromb Vasc Biol,* 2004 24,288-293.

[19] Yoon CH, Hur J, Park KW, Kim JH, Lee CS, Oh IY, Kim TY, Cho HJ, Kang HJ, Chae IH, Yang HK, Oh BH, Park YB, Kim HS. Synergistic neovascularization by mixed transplantation of early endothelial progenitor cells and late outgrowth endothelial cells: The role of angiogenic cytokines and matrix metalloproteinases. *Circulation,* 2005 112,1618-1627.

[20] Fadini GP, Agostini C, Avogaro A. Autologous stem cell therapy for peripheral arterial disease meta-analysis and systematic review of the literature. *Atherosclerosis,* 2010 209,10-17.

[21] Rolland-Turner M, Goretti E, Bousquenaud M, Leonard F, Nicolas C, Zhang L, Maskali F, Marie PY, Devaux Y, Wagner D. Adenosine stimulates the migration of human endothelial progenitor cells. Role of cxcr4 and microrna-150. *PLoS ONE,* 2013 8,e54135.

[22] Oh BJ, Kim DK, Kim BJ, Yoon KS, Park SG, Park KS, Lee MS, Kim KW, Kim JH. Differences in donor cxcr4 expression levels are correlated with functional capacity and therapeutic outcome of angiogenic treatment with endothelial colony forming cells. *Biochem Biophys Res Commun,* 2010 398,627-633.

[23] Heinrich EM, Dimmeler S. Micrornas and stem cells: Control of pluripotency, reprogramming, and lineage commitment. *Circ Res*, 2012 110,1014-1022.

[24] Goretti E, Rolland-Turner M, Léonard F, Zhang L, Wagner DR, Devaux Y. Microrna-16 affects key functions of human endothelial progenitor cells. *J Leukoc Biol*, 2013 93,645-655.

[25] Spinetti G, Fortunato O, Caporali A, Shantikumar S, Marchetti M, Meloni M, Descamps B, Floris I, Sangalli E, Vono R, Faglia E, Specchia C, Pintus G, Madeddu P, Emanueli C. Microrna-15a and microrna-16 impair human circulating proangiogenic cell functions and are increased in the proangiogenic cells and serum of patients with critical limb ischemia. *Circ Res*, 2013 112,335-346.

[26] Fleissner F, Jazbutyte V, Fiedler J, Gupta SK, Yin X, Xu Q, Galuppo P, Kneitz S, Mayr M, Ertl G, Bauersachs J, Thum T. Asymmetric dimethylarginine impairs angiogenic progenitor cell function in patients with coronary artery disease through a microrna-21-dependent mechanism. *Circ Res*, 2010 107,138-143.

[27] Meng S, Cao J, Wang L, Zhou Q, Li Y, Shen C, Zhang X, Wang C. Microrna 107 partly inhibits endothelial progenitor cells differentiation via hif-1beta. *PLoS ONE*, 2012 7,e40323.

[28] Tano N, Kim HW, Ashraf M. Microrna-150 regulates mobilization and migration of bone marrow-derived mononuclear cells by targeting cxcr4. *PLoS ONE*, 2011 6,e23114.

[29] Devaux Y, Vausort M, McCann GP, Zangrando J, Kelly D, Razvi N, Zhang L, Ng LL, Wagner DR, Squire IB. Microrna-150: A novel marker of left ventricular remodeling after acute myocardial infarction. *Circ Cardiovasc Genet*, 2013 April 1.

[30] Minami Y, Satoh M, Maesawa C, Takahashi Y, Tabuchi T, Itoh T, Nakamura M. Effect of atorvastatin on microrna 221 / 222 expression in endothelial progenitor cells obtained from patients with coronary artery disease. *Eur J Clin Invest*, 2009 39,359-367.

[31] Zhang X, Mao H, Chen JY, Wen S, Li D, Ye M, Lv Z. Increased expression of microrna-221 inhibits pak1 in endothelial progenitor cells and impairs its function via c-raf/mek/erk pathway. *Biochem Biophys Res Commun*, 2013 431,404-408.

[32] Zhao T, Li J, Chen AF. Microrna-34a induces endothelial progenitor cell senescence and impedes its angiogenesis via suppressing silent information regulator 1. *Am J Physiol Endocrinol Metab*, 2010 299,E110-116.

[33]  Zhu S, Deng S, Ma Q, Zhang T, Jia C, Zhuo D, Yang F, Wei J, Wang L, Dykxhoorn DM, Hare JM, Goldschmidt-Clermont PJ, Dong C. Microrna-10a* and microrna-21 modulate endothelial progenitor cell senescence via suppressing high-mobility group a2. *Circ Res,* 2013 112,152-164.

[34]  Cheng CC, Lo HH, Huang TS, Cheng YC, Chang ST, Chang SJ, Wang HW. Genetic module and mirnome trait analyses reflect the distinct biological features of endothelial progenitor cells from different anatomic locations. *BMC genomics,* 2012 13,447.

[35]  Zhang Q, Kandic I, Kutryk MJ. Dysregulation of angiogenesis-related micrornas in endothelial progenitor cells from patients with coronary artery disease. *Biochem Biophys Res Commun,* 2011 405,42-46.

[36]  Qiang L, Hong L, Ningfu W, Huaihong C, Jing W. Expression of mir-126 and mir-508-5p in endothelial progenitor cells is associated with the prognosis of chronic heart failure patients. *Int J Cardiol,* 2013, Feb 25.

[37]  Jakob P, Doerries C, Briand S, Mocharla P, Krankel N, Besler C, Mueller M, Manes C, Templin C, Baltes C, Rudin M, Adams H, Wolfrum M, Noll G, Ruschitzka F, Luscher TF, Landmesser U. Loss of angiomir-126 and 130a in angiogenic early outgrowth cells from patients with chronic heart failure: Role for impaired in vivo neovascularization and cardiac repair capacity. *Circulation,* 2012 126,2962-2975.

[38]  van Solingen C, de Boer HC, Bijkerk R, Monge M, van Oeveren-Rietdijk AM, Seghers L, de Vries MR, van der Veer EP, Quax PH, Rabelink TJ, van Zonneveld AJ. Microrna-126 modulates endothelial sdf-1 expression and mobilization of sca-1(+)/lin(-) progenitor cells in ischaemia. *Cardiovasc Res,* 2011 92,449-455.

[39]  Mocharla P, Briand S, Giannotti G, Dorries C, Jakob P, Paneni F, Luscher T, Landmesser U. Angiomir-126 expression and secretion from circulating cd34(+) and cd14(+) pbmcs: Role for proangiogenic effects and alterations in type 2 diabetics. *Blood,* 2013 121,226-236.

[40]  Alaiti MA, Ishikawa M, Masuda H, Simon DI, Jain MK, Asahara T, Costa MA. Up-regulation of mir-210 by vascular endothelial growth factor in ex vivo expanded cd34+ cells enhances cell-mediated angiogenesis. *J Cell Mol Med,* 2012 16,2413 -2421.

[41]  Plummer PN, Freeman R, Taft RJ, Vider J, Sax M, Umer BA, Gao D, Johns C, Mattick JS, Wilton SD, Ferro V, McMillan NA, Swarbrick A, Mittal V, Mellick AS. Micrornas regulate tumor angiogenesis modulated by endothelial progenitor cells. *Cancer Res,* 2013 73,341-352.

[42] Cantaluppi V, Biancone L, Figliolini F, Beltramo S, Medica D, Deregibus MC, Galimi F, Romagnoli R, Salizzoni M, Tetta C, Segoloni GP, Camussi G. Microvesicles derived from endothelial progenitor cells enhance neoangiogenesis of human pancreatic islets. *Cell Transplant,* 2012 21,1305-1320.

[43] Cantaluppi V, Gatti S, Medica D, Figliolini F, Bruno S, Deregibus MC, Sordi A, Biancone L, Tetta C, Camussi G. Microvesicles derived from endothelial progenitor cells protect the kidney from ischemia-reperfusion injury by microrna-dependent reprogramming of resident renal cells. *Kidney Int,* 2012 82,412-427.

[44] Ranghino A, Cantaluppi V, Grange C, Vitillo L, Fop F, Biancone L, Deregibus MC, Tetta C, Segoloni GP, Camussi G. Endothelial progenitor cell-derived microvesicles improve neovascularization in a murine model of hindlimb ischemia. *Int J Immunopathol Pharmacol, 2012* 25,75-85.

In: Progenitor Cells
Editors: P. M. Horton, B. E. Lawrence
ISBN: 978-1-62808-994-3
© 2013 Nova Science Publishers, Inc.

*Chapter 7*

# Cellular Origins in Amphibian Regeneration

*Hai-yan Leng[1], Gufa Lin[2]*, and Ying Chen[2]†*
[1]Department of Hematology, Huashan Hospital, Fudan University, Shanghai, China
[2]Stem Cell Institute, University of Minnesota, Minneapolis, Minnesota, US

## Abstract

Vertebrate animals utilize a diversity of ways to produce cells for regenerating missing body part. In addition to dedifferentiation and transdifferentiation, recent studies in amphibian appendage regeneration have demonstrated the importance of tissue specific progenitor cells in reconstitution of regenerated tissues after amputation or injury. Accumulating evidence supports the notion that a progenitor-based tissue regeneration mechanism is common in vertebrate animals, including mammals. In this chapter, we will examine the cellular origin during amphibian regeneration, and use *Xenopus* limb as an example to discuss the usefulness of progenitor cells for possible measures in stimulating regeneration.

* E-mail: lingufa@gmail.com.
† E-mail: chenx487@umn.edu.

**Keywords:** Regeneration, progenitor cells, lineage restriction, transplantation, stimulating regeneration

# Introduction

The ultimate goal of regenerative medicine is to provoke regeneration of missing or injured body parts. While mammals including human can not regenerate after severe injuries, the urodele amphibian salamanders such as newts and axolotls are capable of regenerating multiple tissue types including limb, tail, heart, jaw, brain and retina, even as adults [1, 2]. The regenerated counterpart is almost a perfect replica, both morphologically and functionally. In this regard, these animals are awarded regeneration champions. Besides the urodeles, anuran amphibians, such as the African clawed frog, *Xenopus laevis*, also possesses remarkable regeneration ability to replace its tail, lens and limb after removal, although the regeneration abilities undergo ontogenetic decline as the animal matures [3, 4]. Researchers have been endeavouring in understanding the underlying cellular and molecular mechanisms of regeneration in these animals. One particularly important aspect is the cellular origin of replaced tissue, as there is increasing interest in regenerative medicine, which often involves use of stem/progenitor cells. With the development of new cell tracing methods, especially the adaption of genetic cell labelling techniques, our understanding of the cellular origins of regenerated tissue is increasingly expanding.

So far there are at least three possible ways by which new cells are produced. First, regenerated tissues could be derived from dedifferentiated cells near the injury site. Upon injury, mature differentiated cells revert to a less differentiated status, re-enter S phase and then redifferentiate to build up new tissue types. Second, transdifferentiation, which means direct conversion of one differentiated cell type to another regardless of cell division, could be a way to replace cells. Last, new cells could arise from resident stem cells or lineage specific progenitor cells. Interestingly, these three routes to regeneration are not species dependent, and some routes seem to be conserved during repair of same tissue type in different animal models. In this chapter, we review our current understanding of these three routes in the classic urodele and anuran amphibian regeneration models. Knowledge obtained from studies on these models may provide clues to develop strategies for therapeutic transplantation in regenerative medicine.

# 1. Dedifferentiation

Dedifferentiation was first carefully proposed in muscle regeneration during salamander forelimb regeneration in 1938 [5]. Later detailed ultrastructure studies confirmed that after amputation, multinucleated myofibers at the limb amputation level broke off, with their nuclei enlarged, and then mononucleate cells were budded off into limb blastema, which was once thought to be composed of a homogeneous dedifferentiated cell population [6]. Some myofiber labelling studies in the 1960s suggested that cells dedifferentiated from muscle fibers near the amputation site were the source of new muscle [7, 8]. However, because of lacking reliable labelling techniques and limitation of photographing equipment, the notion of muscle dedifferentiation to form new myofibers was not readily accepted. Meanwhile, the identification of muscle satellite cells, a type of muscle stem cells, added more complexity in the issue of cell origin of new muscle [9].

More convincing evidence of muscle dedifferentiation during appendage regeneration emerged with better labelling methods. Brockes' lab first re-examined muscle regeneration in newts by grafting rhodamine dextran labelled myotubes into a regenerating limb blastema. After 7 days, they found the labelling dye in mononucleate cells and the number of labelled cells was increased, suggesting proliferation of dedifferentiated muscle cells [10]. To exclude the possibility of cell membrane transfer of labelling dye to other cells in blastema, his lab repeated this experiment by permanently labelling multinucleated myotubes using a genome integrated retrovirus. After transplantation into limb blastema, the myotube gave rise to labelled mononucleate cells, and these cells re-entered cell cycle as shown by BrdU incorporation [11]. These experiments showed that muscle dedifferentiation does occur during muscle regeneration in urodele limbs. Muscle dedifferentiation was also observed in amputated axolotl tails. Echeverri et al. showed that muscle fibers near the amputation plane dedifferentiated and form mononucleate cells that accounted for 17% of the tail blastema cells [12].

During muscle dedifferentiation, cell cycle re-entry occurs before myofiber fragmentation into mononucleate cells [11]. What triggers the cell cycle re-entry? Newt myotubes in culture underwent S-phase upon serum stimulation, with inactivation of the Retinoblastoma (Rb) protein via phosphorylation [13]. Rb is a member of the pocket protein family and is a key negative regulator of G1 to S transition by inhibiting E2F transcription factor [14]. Tanaka et al. further identified that thrombin proteolysis regulated dedifferentiation. A ligand generated downstream from thrombin activation acted directly on newt myotubes to induce cell cycle re-entry [15].

Another important player in muscle dedifferentiation is a homeobox gene called *msx1*, which can induce mammalian myotube dedifferentiation in culture dishes [16]. Kumar et al. examined the relation between newt muscle fiber fragmentation and the expression of *msx1* gene [17]. They found a high correlation between *msx1* expression and the fragmentation of muscle in cultured newt myotubes. However, in the regenerating axolotl tail, upregulation of *msx1* has not been detected in myofibers next to the amputation plane. And neither overexpression nor downregulation of *msx1* appears to affect the course of myofiber dedifferentiation *in vivo* [18].

Although the above experiments all support the concept that dedifferentiated muscle is the source of new muscle in urodeles, a more recent study showed that activation of resident muscle stem cells, the muscle satellite cells, was also involved in newt limb muscle regeneration [19]. This suggests that two, or even more pathways may be utilized in the same tissue reconstitution process, which also provides more options for scientists to develop therapeutic transplantation approaches.

# 2. Transdifferentiation

Transdifferentiation, a subset of metaplasia, means conversion of one differentiated cell type to another, with or without intervening cell division [20]. It used to be generally accepted that fully differentiated cell state is irreversible. But cellular plasticity does exist during regeneration.

Classic examples of transdifferentiation are lens and retina regeneration in amphibians. Wolff first documented that regenerated lens in newts was derived from transdifferentiation of the pigmented epithelial cells of the dorsal iris [21]. During lens regeneration, the pigmented epithelial cells first dedifferentiate to re-enter cell cycle, providing a pool of cells to form lens vesicle for differentiation. Lens differentiates in a process remarkably similar to lens development in terms of sequential appearance of the different crystallins [22, 23]. In contrast to the well-known Wolffian lens regeneration, lens regeneration in anurans, such as *Xenopus laevis*, occurs by transdifferentiation of inner layer cells of cornea to lens cells [24].

Another transdifferentiation phenomenon is retina regeneration in newts. After damage, the retina pigment epithelium loses its original phenotype and re-enters cell cycle to form a neuroepithelial cell layer that eventually differentiate into all the different cell types of retina. Like lens regeneration, newt retina regeneration via transdifferentiation also occurs in adults [25].

Transdifferentiation could also mean switching cell types between germ layers, which violates our understanding of normal development. With transit labelling of single radial glia cell, Echeverri and Tanaka observed that during axolotl tail regeneration ectodermal cell could transdifferentiate to muscle and cartilage, two cell types of mesodermal origin [26]. However, compared to lens and retina regeneration, this event just occurred in a very small amount of cells. And this result needs to be further confirmed by modern genetic labelling techniques.

Since natural transdifferentiation is possible during tissue repair, researchers have been working to decipher mechanisms of triggering transdifferentiation, hoping to directly convert cell types for clinical applications. Several candidates have been identified in the lens and retina transdifferentiation. Fibroblast growth factor 1 (Fgf1) were reported to be an important factor to induce lens regeneration both in newts and frogs [27, 28]. Moreover, Imoka and Brockes demonstrated that damage induced activation of thrombin in dorsal iris was a critical determinant for newt lens regeneration, as inhibition of thrombin activity blocked cell cycle re-entry [29]. Interesting, thrombin is not activated in the iris of the related salamander, axolotl, which cannot regenerate lens. Now there have been many reports of direct conversion of cells with defined factors, some involving trans-germ layer conversions [30]. For example, with overexpression of a small number of factors, liver and pancreas cells can be converted to one and another, and within the pancreas exocrine cells can be converted into endocrine cells [31-33]. Functional neurons can be directly converted from fibroblast [34]. These approaches to induce transdifferentiation bear great promise in regenerative medicine.

# 3. Progenitor Cells

The advancement of transgenic method that permanently labels specific tissue type with genome-integrated reporters such as green fluorescent protein (GFP) allows investigation of tissue origins during appendage regeneration with high precision. Recent lineage tracing experiments in regeneration of the frog tail and the axolotl limb have demonstrated that lineage restricted stem/progenitor cells are the main source for new cells in appendage regeneration.

The first detailed lineage tracing analysis of appendage regeneration in a vertebrate was done in *Xenopus* tadpoles. *Xenopus* tadpoles can fully regenerate its tail after amputation. The tadpole tail has well-demarcated, distinct tissue types. It contains a notochord in the centre, a spinal cord in the dorsal midline, and segmented myotome on the sides. These three main axial structures are surrounded with connective tissues and epidermis. The cellular origin in *Xenopus* tadpole tail regeneration was thoroughly analysed by tissue grafts to generate animals with one of the three major tissues (spinal cord, notochord and muscle) genetically labelled with GFP. The *CMV* promoter-driven GFP expression in donors remains active all the time, so after the tail is amputated through the graft, a labelled cell should still express GFP even if it dedifferentiates or redifferentiates into the same or a different tissue type. The cell lineage of each tissue in the tail can then be traced during regeneration.

Of particular interest is whether there is dedifferentiation of mature multinucleate myofibers during muscle regeneration in *Xenopus*, as fragmentation of myofibers had been well documented in urodele tail muscle regeneration. To label muscle tissue in the tadpole tail, a piece of presomitic mesoderm was transplanted in early or late neurula. This graft could label the myofibers in the tadpole tail. However after amputation, myofibers in the regenerates were rarely labelled with GFP from early neurulae grafts. But if the grafts were taken from a more lateral region of the presomitic mesoderm, or from a late stage neurulae in which the lateral tissue has moved more dorsally, the regenerating tails contained significant number of GFP labelled myofibers [35, 36] (Figure 1A). This result firstly suggests that there is no dedifferentiation in *Xenopus* tail muscle regeneration; Otherwise, GFP expression would be found in newly formed myofibers from amputations through the region containing labelled myofibers. Secondly, it indicates that some other type of cells is the precursors for the regenerated muscle in the tadpole tail. Further investigation showed that, as in mammalian muscle regeneration, Pax7 expressing muscle satellite cells, which were included in the grafts of more lateral region of the presomitic mesoderm, are the source of new muscles in *Xenopus* tadpole tail regenerates [35, 37]. Satellite cells are small mononuclear cells lying within the basement membrane of the myofibers [9]. Upon injury, satellite cells are activated to proliferate and differentiate to fuse into muscle fibers [38]. The regenerating myofiber was only labelled when satellite cells were labelled in the tadpole tail (Figure 1B). When the satellite cell pool is depleted, such as by loss of Pax7 function, muscle regeneration in the tadpole tail was severely reduced [37]. So unlike urodeles,

*Xenopus* tadpoles regenerate their muscle via activation of muscle stem cells, rather than through muscle dedifferentiation.

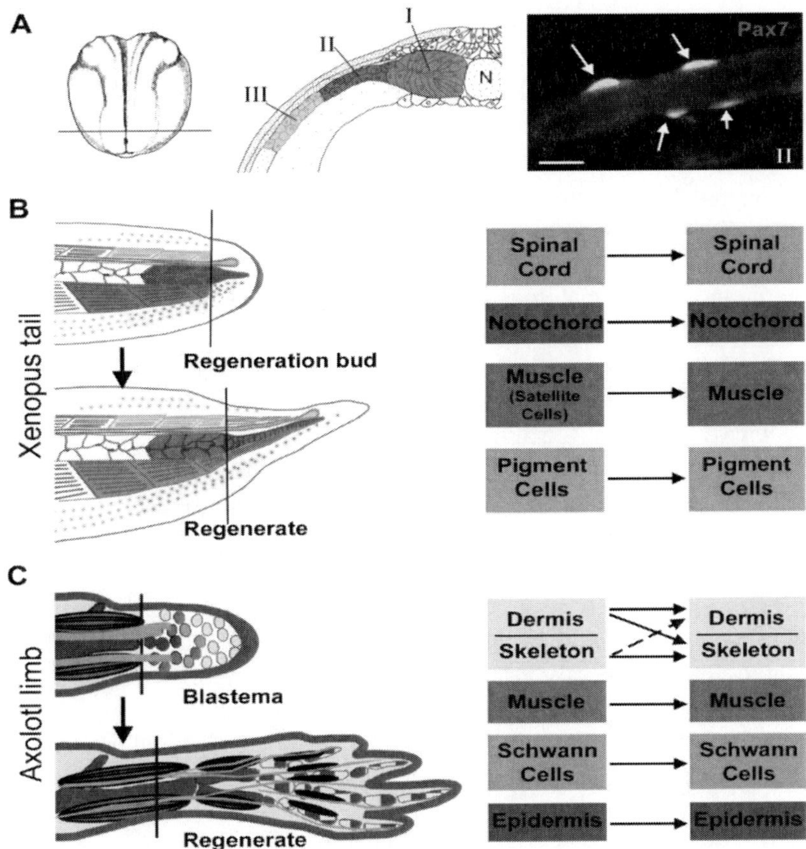



Figure 1. Cell origins in tail and limb regeneration. **A**. Labeling of muscle satellite cells that regenerate new myofibers in the *Xenopus* tadpole tail. The lateral mesoderm (II) of early neurulae embryos gives rise to satellite cells in muscle fibers (right panel, arrows, stained for pax7). Paraxial mesoderm (I) or ventral mesoderm (III) labels little satellite cells. N, notochord. **B**. *Xenopus* tail regeneration. Each major tissue type in the tadpole tail regenerates from its own corresponding tissue in the stump. Lines indicate amputation level. **C**. Limb regeneration in axolotl. The limb blastema consists of progenitor cells that remember their origin in the stump and know their destination in the regenerate. Lines indicate amputation level. A adapted with permission from Daughters et al. (2011) Development; C adapted from Kragl et al. (2011) Nature, with permission.

To follow regeneration of the spinal cord, a portion of the spinal cord in the tadpole tail was labelled either by embryonic graft of posterior neural plate from *CMVGFP* transgenic neurula embryo to wild type host [35], or by replacement of a piece of spinal cord with *CMVGFP* transgenic spinal cord at tadpole stages [39]. After amputation, it was found that as long as 0.5 mm of spinal cord was labelled in the stump, the whole regenerating spinal cord was GFP labelled, suggesting that this length of spinal cord provides all the cells needed for reforming new spinal cord. The spinal cord closes off and forms a neural ampulla with the proliferation of progenitor cells. A recent study suggests that cells expressing the transcription factor Sox2 maybe the progenitors of regenerated spinal cords. Sox2 expressing cells were found to locate in the ependymal region of the spinal cord and accumulate in the ablation site after spinal cord transection. The expression level of Sox2 seems to correlate to the regeneration capacity of spinal cord as down-regulation of Sox2 inhibits regeneration [40]. In the axolotl tail it has recently been shown that clonally generated neurospheres from single spinal cord cell could integrate into the spinal cord after transplantation, and upon tail amputation the transplanted neurosphere had the potential to generate all cell types in the regenerating spinal cord, suggesting that there is a cell type with the property of neural stem cells in the spinal cord [41].

Regeneration of the notochord in the tadpole tail seems to be very straightforward. When the notochord was labelled and amputated, new notochord formed directly from the proximal notochord, forming a bullet–shaped mass of cells that are continuous with the sheath of the more proximal region. GFP was found only in the regenerating notochord. Other tissue types in the regenerating tail, such as the skin and pigment cells, are also derived from their corresponding tissue types in the stump [39].

The above analyses show that tissues in the tadpole tail (the spinal cord, the notochord and the myofibers) behave independently during regeneration. Each tissue type regenerates from their respective specialized precursors (Figure 1B). There is no dedifferentiation of mature muscle fibers and there is no metaplasia involved in *Xenopus* tail regeneration.

Using the same cell labeling strategy in the tadpole tail system, Tanaka's group obtained similar observations in axolotl limb regeneration [42]. Kragl et al. showed that during axolotl limb regeneration, labelled epidermis becomes only epidermis. Muscle becomes only new muscle. GFP labeled Schwann cells become only new Schwann cells. GFP labelled dermis makes new dermis, cartilage and tendons, tissues that share a common lateral mesoderm origin in development. So in agreement with the *Xenopus* tail experiment, the results show that the axolotl limb blastema cells are not a homogeneous undifferentiated cell population as previously thought. Instead the limb blastema in axolotl consists of a heterogeneous pool of lineage restricted progenitor cells that do not switch between embryonic germ layers during regeneration (Figure 1C).

Taken together, the observations in amphibian appendage regeneration demonstrate that in addition to dedifferentiation and transdifferentiation, lineage specific stem/progenitor cells play very important roles in amphibian appendage regeneration.

# 4. Stimulating Limb Regeneration by Progenitor Cells

The limb of *Xenopus* is an ideal model for gain-of-function studies and for measures to stimulate regeneration. This is because of its interesting regenerative behavior. The *Xenopus* tadpole normally regenerates its limbs at early developmental stages when the limb bud is a flattened paddled shape. But this ability is gradually lost in late stage tadpoles as the limb bud differentiates. In post-metamorphic frogs the limbs do not regenerate except producing an unsegmented cartilaginous spike after amputation [43]. The unique regeneration behavior of the *Xenopus* limbs has prompted many attempts to stimulate regeneration in the non-regenerating postmetamorphic frogs. Some methods may have worked in tadpole limbs, but most claims of stimulated regeneration in postmetamorphic frog limbs have proved irreproducible [44, 45].

As described in previous sections, appendage regeneration relies heavily on the proliferation of progenitor cells. Early studies on *Xenopus* limb regeneration also showed that regeneration capacity is an intrinsic property of the developing limb [46, 47]. This suggests that limb progenitor cells may be used to improve regeneration. Lin et al. tested this idea by grafting limb progenitor cells isolated from regeneration competent tadpoles to the non-regenerating frog forelimb stump and showed that the larval limb progenitor cells do stimulate limb regeneration [48] (Figure 2).

The success requires that the limb progenitor cells to be high in Wnt/beta-catenin signaling activity, and they are supplemented with extracellular factors such as Shh and Fgfs. More importantly, the cells need to be delivered in a manner that enables good cell survival, such as by mixing the cells in a fibrin matrix. Although the regenerated limbs are far from normal, they contain multiple digits and the digits are segmented, heavily innervated and in some cases also express the joint marker *gdf5*, suggesting the possible formation of joints in these regenerates. Consistent to the idea that cells have memories of their developmental origin, only progenitor cells isolated from early tadpole limbs have the capacity to stimulate regeneration in the adult frog limb. Regeneration competent cells from tadpole tail regeneration bud do not have the ability to promote regeneration, even with elevated expression of Wnt/beta-catenin and with the same treatment. Notably, addition of Shh and Fgf10 supports the survival and proliferation of transplanted progenitor cells, but application of Shh, Fgf10 beads in limbs without progenitor cell transplantation does not promote regeneration [48]. These observations demonstrate the critical requirement for progenitor cells in stimulating limb regeneration.

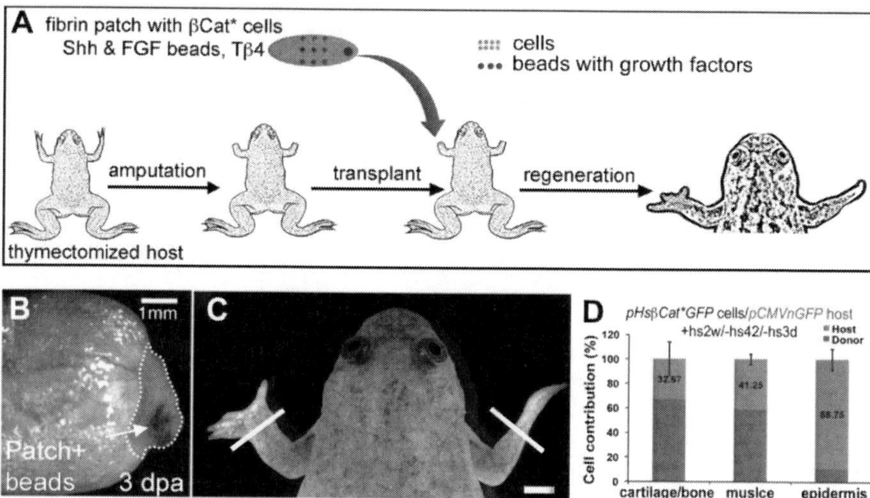



Figure 2. Stimulating limb regeneration with progenitor cell transplantation. A. An approach to stimulate limb regeneration in adult frogs, involving transplantation of limb progenitor cells and application of exogenous factors to limb stump in frogs thymectomized at tadpole stages to reduce immune response. B. Transplantation of cells in a fibrin patch (dotted lines) onto limb stump. Beads containing growth factors (arrow) are implanted in the fibrin patch. C. A frog with stimulated limb regeneration (left). The right limb is untreated control. Lines indicate amputation levels. D. Host limb cells are mobilized to contribute limb regeneration. Adapted from Lin et al. (2013) Dev. Cell, with permission.

Remarkably, the regenerates after progenitor cell transplantation not only consist of donor cells, but also consist of host cells, with a substantial contribution of host cells in the regenerated cartilage and muscle. This suggests that the transplanted progenitor cells not only participate in the regeneration of new tissues, but also recruit and mobilize the host cells, despite their lack of regeneration competence, to participate in the regeneration process.

# Conclusion and Perspectives

Recent studies of the cellular basis in various regeneration model systems reveal that injured tissues use a diversity of ways to provide progenitor cells for regeneration. Regeneration of lens and retina occurs through transdifferentiation. Regeneration of appendages, such as limbs and tails, employs lineage restricted progenitor/stem cells. Both dedifferentiation and activation of stem cells may contribute to the proliferating progenitors for regeneration. This is also true for other vertebrate appendage regeneration systems such as the fish fin and mouse digit tip. In the fish caudal fin, cell-tracing analyses demonstrated that cells in the fin blastema have lineage restrictions and do not contribute to other cell types [49, 50]. Differentiated osteoblasts can dedifferentiate and enter into blastema, but the dedifferentiated osteoblasts mainly contribute to new osteoblasts. Detailed lineage tracing analysis of digit tip regeneration in the mouse showed that like the frog tail and the salamander limb, the blastema that regenerate the digit tip consists of lineage-restricted progenitor cells [51, 52].

These studies have significant implications for regenerative medicine. Successful stimulation of limb regeneration in the non-regenerating frog by limb progenitor cells and reconstitution of the axolotl spinal cord by neurospheres are excellent examples of using progenitor cells for therapeutic treatment. It is of great interest to test whether similar transplantation of progenitor cells will promote regeneration in mammals. It is noteworthy that the behavior of transplanted progenitor cells differs greatly in amphibians and mammals. For example, in axolotl there is an extensive neurogenesis and reconstruction of all spinal cord cell types after transplantation of neurospheres [41], while mammalian neurosphere transplantation often results in astrocyte formation, and limited neurogenesis [53, 54]. Understanding of the underlying mechanism of the difference should aid in the development of regenerative strategies in spinal cord injury. Thus, the amphibian regeneration models could serve as an excellent platform for further characterization of the cellular and molecular mechanisms involved in successful regeneration and for optimization of protocols that can be applied to humans.

# References


[1] Brockes, J. P. and Kumar, A. (2008). Comparative Aspects of Animal Regeneration. *Annu. Rev. Cell Dev. Biol.* 24, 525-549.

[2] Nacu, E. and Tanaka, E. M. (2011). Limb regeneration: a new development? *Annu. Rev. Cell Dev. Biol.* 27, 409-40.

[3] Beck, C. W., Izpisua Belmonte, J. C. and Christen, B. (2009). Beyond early development: Xenopus as an emerging model for the study of regenerative mechanisms. *Dev. Dyn.* 238, 1226-48.

[4] Slack, J. M., Lin, G. and Chen, Y. (2008). The Xenopus tadpole: a new model for regeneration research. *Cell Mol. Life Sci.* 65, 54-63.

[5] Thornton, C. S. (1938). The histogenesis of muscle in the regenerating forelimb of larval Ambystoma punctatum. *J. Morphol.* 62, 219-235.

[6] Hay, E. D. (1959). Electron microscopic observations of muscle dedifferentiation in regenerating Amblystoma limbs. *Dev. Biol.* 1, 555-585.

[7] Hay, E. D. and Fischman, D. A. (1961). Origin of the blastema in regenerating limbs of the newt Triturus viridescens. An autoradiographic study using tritiated thymidine to follow cell proliferation and migration. *Dev. Biol.* 3, 26-59.

[8] Steen, T. P. (1968). Stability of chondrocyte differentiation and contribution of muscle to cartilage during limb regeneration in the axolotl (Siredon mexicanum). *J. Exp. Zool.* 167, 49-78.

[9] Mauro, A. (1961). Satellite cell of skeletal muscle fibers. *J. Biophys. Biochem. Cytol.* 9, 493-5.

[10] Lo, D. C., Allen, F. and Brockes, J. P. (1993). Reversal of muscle differentiation during urodele limb regeneration. *Proc. Natl. Acad. Sci. U.S.A.* 90, 7230-4.

[11] Kumar, A., Velloso, C. P., Imokawa, Y. and Brockes, J. P. (2000). Plasticity of retrovirus-labelled myotubes in the newt limb regeneration blastema. *Dev. Biol.* 218, 125-136.

[12] Echeverri, K., Clarke, J. D. W. and Tanaka, E. M. (2001). In vivo imaging indicates muscle fiber dedifferentiation is a major contributor to the regenerating tail blastema. *Dev. Biol.* 236, 151-164.

[13] Tanaka, E. M., Gann, A. A., Gates, P. B. and Brockes, J. P. (1997). Newt myotubes reenter the cell cycle by phosphorylation of the retinoblastoma protein. *J. Cell Biol.* 136, 155-65.

[14] Burkhart, D. L. and Sage, J. (2008). Cellular mechanisms of tumour suppression by the retinoblastoma gene. *Nat. Rev. Cancer* 8, 671-82.

[15] Tanaka, E. M., Drechsel, D. N. and Brockes, J. P. (1999). Thrombin regulates S-phase re-entry by cultured newt myotubes. *Curr. Biol.* 9, 792-9.

[16] Odelberg, S. J., Kollhoff, A. and Keating, M. T. (2000). Dedifferentiation of mammalian myotubes induced by msx1. *Cell* 103, 1099-109.

[17] Kumar, A., Velloso, C. P., Imokawa, Y. and Brockes, J. P. (2004). The regenerative plasticity of isolated urodele myofibers and its dependence on MSX1. *PLoS Biol.* 2, E218.

[18] Schnapp, E., Kragl, M., Rubin, L. and Tanaka, E. M. (2005). Hedgehog signaling controls dorsoventral patterning, blastema cell proliferation and cartilage induction during axolotl tail regeneration. *Development* 132, 3243-53.

[19] Morrison, J. I., Loof, S., He, P. and Simon, A. (2006). Salamander limb regeneration involves the activation of a multipotent skeletal muscle satellite cell population. *J. Cell Biol.* 172, 433-40.

[20] Slack, J. M. (2007). Metaplasia and transdifferentiation: from pure biology to the clinic. *Nat. Rev. Mol. Cell Biol.* 8, 369-78.

[21] Wolff, G. (1895). Entwicklungsphysiologische Studien I. Die Regeneration der Urodelenlinse. *Wilhelm Roux's Arch. Entw. Mechan. Org.* 1, 380-390.

[22] Yamada, T. (1977). Control mechanisms in cell-type conversion in newt lens regeneration. *Monogr. Dev. Biol.* 13, 1-126.

[23] Mizuno, N., Agata, K., Sawada, K., Mochii, M. and Eguchi, G. (2002). Expression of crystallin genes in embryonic and regenerating newt lenses. *Dev. Growth Differ.* 44, 251-6.

[24] Freeman, G. (1963). Lens Regeneration from the Cornea in Xenopus Laevis. *J. Exp. Zool.* 154, 39-65.

[25] Tsonis, P. A. and Del Rio-Tsonis, K. (2004). Lens and retina regeneration: transdifferentiation, stem cells and clinical applications. *Exp. Eye Res.* 78, 161-72.

[26] Echeverri, K. and Tanaka, E. M. (2002). Ectoderm to mesoderm lineage switching during axolotl tail regeneration. *Science* 298, 1993-1996.

[27] Hyuga, M., Kodama, R. and Eguchi, G. (1993). Basic fibroblast growth factor as one of the essential factors regulating lens transdifferentiation of pigmented epithelial cells. *Int. J. Dev. Biol.* 37, 319-26.

[28] Bosco, L., Testa, O., Venturini, G. and Willems, D. (1997). Lens fibre transdifferentiation in cultured larval Xenopus laevis outer cornea under the influence of neural retina-conditioned medium. *Cell Mol. Life Sci.* 53, 921-8.

[29] Imokawa, Y. and Brockes, J. P. (2003). Selective activation of thrombin is a critical determinant for vertebrate lens regeneration. *Curr. Biol.* 13, 877-81.

[30] Ladewig, J., Koch, P. and Brustle, O. (2013). Leveling Waddington: the emergence of direct programming and the loss of cell fate hierarchies. *Nat. Rev. Mol. Cell Biol.* 14, 225-36.

[31] Shen, C. N., Slack, J. M. and Tosh, D. (2000). Molecular basis of transdifferentiation of pancreas to liver. *Nat. Cell Biol.* 2, 879-87.

[32] Horb, M. E., Shen, C. N., Tosh, D. and Slack, J. M. (2003). Experimental conversion of liver to pancreas. *Curr. Biol.* 13, 105-15.

[33] Zhou, Q., Brown, J., Kanarek, A., Rajagopal, J. and Melton, D.A. (2008). In vivo reprogramming of adult pancreatic exocrine cells to beta-cells. *Nature* 455, 627-32.

[34] Vierbuchen, T., Ostermeier, A., Pang, Z. P., Kokubu, Y., Sudhof, T. C. and Wernig, M. (2010). Direct conversion of fibroblasts to functional neurons by defined factors. *Nature* 463, 1035-41.

[35] Gargioli, C. and Slack, J. M. (2004). Cell lineage tracing during Xenopus tail regeneration. *Development* 131, 2669-79.

[36] Daughters, R. S., Chen, Y. and Slack, J. M. (2011). Origin of muscle satellite cells in the Xenopus embryo. *Development* 138, 821-30.

[37] Chen, Y., Lin, G. and Slack, J. M. W. (2006). Control of muscle regeneration in the Xenopus tadpole tail by Pax7. *Development* 133, 2303-2313.

[38] Le Grand, F. and Rudnicki, M. A. (2007). Skeletal muscle satellite cells and adult myogenesis. *Curr. Opin. Cell Biol.* 19, 628-33.

[39] Lin, G., Chen, Y. and Slack, J. M. (2007). Regeneration of neural crest derivatives in the Xenopus tadpole tail. *BMC Dev. Biol.* 7, 56.

[40] Gaete, M., Munoz, R., Sanchez, N., Tampe, R., Moreno, M., Contreras, E. G., Lee-Liu, D. and Larrain, J. (2012). Spinal cord regeneration in Xenopus tadpoles proceeds through activation of Sox2-positive cells. *Neural Dev.* 7, 13.

[41] McHedlishvili, L. et al. (2012). Reconstitution of the central and peripheral nervous system during salamander tail regeneration. *Proc. Natl. Acad. Sci. U.S.A.* 109, E2258-66.

[42] Kragl, M., Knapp, D., Nacu, E., Khattak, S., Maden, M., Epperlein, H.H. and Tanaka, E.M. (2009). Cells keep a memory of their tissue origin during axolotl limb regeneration. *Nature* 460, 60-5.

[43] Dent, J. N. (1962). Limb regeneration in larvae and metamorphosing individuals of the South African clawed toad. *J. Morphol.* 110, 61-77.

[44] Muller, T. L., Ngo-Muller, V., Reginelli, A., Taylor, G., Anderson, R. and Muneoka, K. (1999). Regeneration in higher vertebrates: Limb buds and digit tips. *Semin. Cell Dev. Biol.* 10, 405-413.

[45] Carlson, B. M. (2007) Principles of Regenerative Biology, Academic Press. Burlington MA.

[46] Muneoka, K., Hollerdinsmore, G. and Bryant, S. V. (1986). Intrinsic Control of Regenerative Loss in Xenopus-Laevis Limbs. *J. Exp. Zool.* 240, 47-54.

[47] Sessions, S. K. and Bryant, S. V. (1988). Evidence that regenerative ability is an intrinsic property of limb cells in Xenopus. *J. Exp. Zool.* 247, 39-44.

[48] Lin, G., Chen, Y. and Slack, J. M. W. (2013). Imparting Regenerative Capacity to Limbs by Progenitor Cell Transplantation. *Dev. Cell* 24, 41-51.

[49] Tu, S. and Johnson, S. L. (2011). Fate restriction in the growing and regenerating zebrafish fin. *Dev. Cell* 20, 725-32.

[50] Knopf, F. et al. (2011). Bone regenerates via dedifferentiation of osteoblasts in the zebrafish fin. *Dev. Cell* 20, 713-24.

[51] Rinkevich, Y., Lindau, P., Ueno, H., Longaker, M. T. and Weissman, I. L. (2011). Germ-layer and lineage-restricted stem/progenitors regenerate the mouse digit tip. *Nature* 476, 409-13.

[52] Lehoczky, J. A., Robert, B. and Tabin, C. J. (2011). Mouse digit tip regeneration is mediated by fate-restricted progenitor cells. *Proc. Natl. Acad. Sci. U.S.A.* 108, 20609-14.

[53] Akesson, E., Piao, J. H., Samuelsson, E. B., Holmberg, L., Kjaeldgaard, A., Falci, S., Sundstrom, E. and Seiger, A. (2007). Long-term culture and neuronal survival after intraspinal transplantation of human spinal cord-derived neurospheres. *Physiol. Behav.* 92, 60-6.

[54] Wu, S., Suzuki, Y., Kitada, M., Kitaura, M., Kataoka, K., Takahashi, J., Ide, C. and Nishimura, Y. (2001). Migration, integration, and

differentiation of hippocampus-derived neurosphere cells after transplantation into injured rat spinal cord. *Neurosci. Lett.* 312, 173-6.

# Index

## D

## J



## K



## L

## O



## P

## T